Applications of Liapunov Methods in Stability

Mathematics and Its Applications

Volume 245

Applications of Liapunov Methods in Stability

by

A. Halanay

Faculty of Mathematics,
University of Bucharest,
Bucharest, Romania

and

V. Răsvan

Faculty of Control and Computer Engineering,
University of Craiova,
Craiova, Romania

SPRINGER SCIENCE+BUSINESS MEDIA, B.V.

Library of Congress Cataloging-in-Publication Data

Halanay, Aristide.
Applications of Liapunov methods in stability / by A. Halanay and V. Răsvan.
p. cm. -- (Mathematics and its applications ; v. 245)
Includes index.
ISBN 978-94-010-4697-8 ISBN 978-94-011-1600-8 (eBook)
DOI 10.1007/978-94-011-1600-8
1. Lyapunov functions. 2. Stability. I. Răsvan, V. II. Title.
III. Series: Mathematics and its applications (Kluwer Academic Publishers) ; v. 245.
QA871.H188 1993
003'.01'51535--dc20 92-43812

ISBN 978-94-010-4697-8

Printed on acid-free paper

Contents

Preface

The year 1992 marks the centennial anniversary of publication of the celebrated monograph "The General Problem of Stability of Motion" written by A.M.Liapunov. This anniversary inspires to think about the way theory and applications have developed during this century. The first observation one can make is that the so-called "second method", nowadays known as the "Liapunov function method", has received more attention than the "first method"; let us also mention the study of critical cases, which brought more attention recently in connection with the study of bifurcations and with nonlinear stabilization. One of the reasons of popularity of the Liapunov function approach might be the fact that, in many situations in science and engineering, and not only in mechanics, which was the main source of inspiration for the work of Liapunov, natural Liapunov functions may be proposed, intimately connected with the properties of the processes. It is one of the purposes of this book to advocate this idea.

From the mathematical viewpoint, the century after the first appearance of Liapunov's monograph has been characterized both by generalizations and by refinements of Liapunov's ideas. But we feel that the most spectacular progress is the understanding of the wide possibilities open for applications by the use of Stability Theory as constructed by Liapunov a century ago.

We have tried to show some of the ideas in this direction by starting with our personal experience in the study of some models. It is extremely inspiring to see that good models in science and engineering have nice properties which allow us to associate natural Liapunov functions and use them to discuss the corresponding stability properties. Such examples have been encountered by us in Power Engineering,

Chemical Engineering and Economic Theory, and the book we propose is the result of our experience in these fields. We are convinced that not only the beauty of final results, the diversity of the mathematical tools which we used, but also the approach itself from models and problems to mathematical analysis might be inspiring for other fields of application.

The authors

About the Notations

The fact that this is an application–oriented book has determined the notations used throughout it. In the part of the book where theory is discussed, the standard notation of the field is used. Among them the well–known convention of representing scalars by lower case Greek letters, vectors by lower case Latin letters and matrices by capital Latin letters has been employed. There are, of course, notable deviations: the independent variable, denoted, as usual, by t, or the scalar Liapunov function denoted by V. Other deviations are either mentioned or clear from the context. In the part concerned with applications this convention is no longer respected: here the priority was granted to the specific notations of the field (Engineering, Economics, Ecology).

Some general notations are as follows. By $\Re\gamma$, $\Im\gamma$, $\overline{\gamma}$ we denoted the real, the imaginary and the conjugate of the complex number γ. If A is a matrix, A^* is the transpose and conjugate, or only the transpose if A has real entries; the same is valid for vectors too. The norm is always denoted by $|\cdot|$; A^{-1} is the inverse of A and $\det A$ is its determinant. By $\mathrm{diag}(\lambda_1, \ldots, \lambda_n)$ we denoted the diagonal matrix with $(\lambda_1, \ldots, \lambda_n)$ as the entries on the main diagonal; by $\mathrm{col}(\xi_1, \ldots, \xi_n)$ we denoted the column vector with the entries $(\xi_1, \ldots, \xi_n)$. The identity matrix is denoted by I, regardless its dimension (which follows from the context); the same is valid for 0. If $f(t)$ is a time function, its Laplace transform is denoted by $\tilde{f}(\sigma)$ and its Fourier transform by $\tilde{f}(\imath\omega)$. The derivative of the time function f is denoted both as $\dot{f}$ and f'.

CHAPTER 1

Introduction

According to the classical definition of Malkin (1952, p.9), stability theory deals with the effect of disturbances on time–processes in real systems.

The stability problems originated from mechanics and this is illustrated, for example, by the fact that, in various fields in science and engineering, time–processes are usually called motions. In mechanics it is realized that not all equilibrium states resulting from equilibrium equations are realized and observed in practice. This can be explained by taking into account "small forces" and "small deviations" from the initial equilibrium state which are neglected in the equilibrium equations. All these factors are disturbances which tend to modify the equilibrium. If these modifications are small and the system remains near the equilibrium state, or even returns back to this state, the state of equilibrium is called stable. On the contrary, if such influences are important and the system tends to depart from a state of equilibrium, then the equilibrium is called unstable.

The same situation can occur for any motion of the system. The motion whose stability is in question is called ***basic*** and, according to the principles of mechanics, is determined mainly by some ***basic forces.*** But there are always some small, negligible factors, not taken into account, which represent disturbances. In fact, this corresponds to the representation of the system as subject to the basic forces and embedded in a low–amplitude disturbing field.

If the disturbances are assumed to be "short–period", their effect can be incorporated in the initial conditions of a new - perturbed - motion. If this new motion remains in a neighbourhood of the basic one, the basic motion is called stable, otherwise it is called unstable. This point of view of incorporating the effect of short–period disturbances in the initial conditions of a perturbed motion showed itself fruitful in other fields of science and engineering, some of which being rather far from Rational Mechanics.

The Stability Problem in Engineering

Practical experience shows that in engineering processes are also subject to some disturbances which tend to modify their basic properties.

For instance, if *nuclear power reactor physics and engineering* is considered, one of the basic requirements is the stability of the steady state with respect to small disturbances. Here, a steady state is a state whose characteristic parameters (neutron and thermal agent flows, temperatures, isotope concentrations, etc.) do not deviate from their values if external disturbances are absent.

In the case of nuclear power reactions, the most usual disturbances are: grid frequency variations, modification of the thermal agent regime, Xe–135 isotope effects, etc. For instance, the interaction of the neutron flow and of Xe–135 isotopes can give an instability of the delivered power. Namely, the increase of the neutron flow (which is a measure of delivered power) can affect the Xenon twicely: first, decrease of the Xe–135 activates the fission, thus leading to further increase of the delivered power; second, an increased neutron flow increases the quantity of I–135 which is converted to Xe–135 via a nuclear reaction having a time constant of 6–7 hours; in this way the fission is slowed down, hence the neutron flow decreases. These two opposite effects, the first of them being instantaneous and the second one delayed, are able to generate large amplitude oscillations (swing–swell) of the reactor power level.

The effects of external phenomena can be sensed in the delivered power via neutron reacting properties. If all properties characterizing reactor's fission reactions are combined under the term reactivity, it is obvious that reactivity is the control parameter of power generation. The reactivity can be modified both by external (temperature, pressure, automatic feedback control) and internal factors, e.g., Xe–135 poisoning.

But some of these factors are themselves influenced by the power generated, hence, we can represent reactor phenomena as a feedback structure. Moreover, from the above consideration we keep in mind the fact that such a feedback structure can produce an instability generated by oscillations of the systems technical parameters.

Concerning the disturbances, it should be mentioned that these can occur in any part of the nuclear power plant. Moreover, if only a subsystem of the plant is a subject of study – e.g., the active zone – then those variables which are specific to the rest of the plant should be considered as disturbances.

Besides these disturbances there are other perturbations that should be taken into account; these are connected with model uncertainties: disintegration constants, average neutron "life" time, and reactivity are in fact disturbances of the system. Moreover, taking into account the complexity of the described phenomena, the disturbances affecting all process variables should be considered as well.

If all these disturbances are regarded as short–period, their effect can be viewed as being included in the initial conditions by performing a judgment of the same type as previously: when any disturbance occurs, the system's evolution obeys a set of laws which are different from those described by the model equations. When the disturbance disappears the system's state is different from equilibrium. If this state is taken as initial, then, in order to obtain stability of the reactor, the system must recover the equilibrium state or at least stay in a neighbourhood of it. It follows that stability must be studied using processes generated by initial conditions which are in some neighbourhood of the equilibrium state.

Perturbation of a reactor's equilibrium point can be also performed in the following way: by modifying the steady–state regime according to the control schedule (e.g., by imposing a new power level). This modification – called operation or manoeuvre – is performed by modifying some system's parameters; in fact, by modifying the system. Therefore, the current stationary point becomes an initial condition for a transient process that should converge to another stationary point. This is another case when the effect of some disturbances is incorporated in the initial conditions.

Finally, the short–period faults also generate disturbances. During a fault the operating conditions for the nuclear reactor are changed and the initial steady–state regime is no longer steady–state. Therefore a transient process starts, and when the fault is removed the reactor has attained some state as a consequence of the transitions. The fault being removed, the reactor must return to the initial steady–state regime. In this way, a transient process generated by initial conditions occurs.

Similar problems can be observed in other fields of engineering. In *chemical engineering* the stability for chemical reactors has to be considered already at the stage of designing the reactor, because its stable operating around some steady–state point has to be ensured both at the laboratory and industrial stages. Chemical kinetics, mass and heat transfer processes, etc. are subject to various disturbances, such as: initial concentrations of the reacting substances, thermodynamic and environmental parameters modifications, mass flow variations, etc. It is obvious that many of these disturbances cannot be described completely, hence, starting from the design stage one has to be sure that "when the disturbing effects cease, the deviations from the steady–state regime must tend to an arbitrarily small value which is preassigned" (R.Aris, 1961).

In this case also one has to consider the evolution of the system in a neighbourhood of an equilibrium state – considered as a basic state – due to some temporarily acting disturbances. If these disturbances are short–period, their effect can be included in the initial conditions.

Chemical reactor operation also requires planned modifications of the steady–state point: in this case the former stationary point becomes an initial condition for a transient process to the new stationary point.

A stability problem occurs also when considering *hydroelectric engineering.* In order to illustrate the ideas, consider the simplest hydroelectric power plant model: it consists of a storage pool and a hydraulic turbine–generator group connected by a head race and a surge tank.

For a given turbine power a steady–state has to be considered. It is described by certain constant parameters such as head race flow pressure losses, surge tank water level, etc.

In order to understand the way of disturbing a stationary state of the plant it is necessary to consider once again the manoeuvre. In hydraulic engineering the manoeuvre is a way of modifying the water flow regime between the storage pool and the turbine.

The most usual manoeuvres are valve opening (for an increase of the generated electrical power) and valve closing (for a decrease of the electrical power). Such a manoeuvre shows a transition from a steady state to another steady state. This transition is accompanied by two basic phenomena: the waterhammer and the level oscillations of the surge tank.

The waterhammer is described by violent (i.e. high level and high frequency) oscillations of flow parameters during the manoeuvre or immediately after it. Waterhammer occurs due to propagation phenomena on the head race as a result of the superposition of the reflected (by the plant valves) and the forward pressure waves. To attenuate the waterhammer (which can result in tube breaking or turbine blades damage) it is necessary to compensate the forward wave by the reflected one. This goal can be achieved if the sign changing reflection of the wave takes place before the reflection in the initial cross–section of the head race. This condition is met on the free cross–section of the surge tank where the water pressure is constant and equals the atmospheric pressure, provided this surge tank is positioned closer to the hydraulic turbine. Due to this fact some of the shock wave energy is dissipated by the oscillations of the surge tank water mass. These oscillations are quite slow, hence the manoeuvre duration can be neglected and the head race flow variation can be assumed to be stepwise.

The decomposition of the transient process into waterhammer and surge tank mass oscillations is nevertheless conventional. In fact, there exists a unique transient process whose phases differ by the influence of various factors during the transitions: some factors have a larger influence during the first phase and their influence is diminished during the second phase, while other factors with a reduced influence in the first phase become important in the second phase.

From what was said above, one can see now what is understood by stability in hydraulics: waterhammer and surge tank mass oscillations damping and the operation of the plant in the new (past–manoeuvre) operating point.

Here also one can see a previously met property: when the system is in steady–state regime, any modification of its characteristics leading to a change of the steady–state acts as a disturbance. Indeed, modification of system's characteristics means that the system's behaviour obeys other laws than before and the resulting steady–state is different from the initial one. This initial steady–state is in fact the "prime mover" of the system's evolution to the steady state required by the new situation. Therefore, hydraulic stability is also described by the system's evolution in a neighbourhood of a certain state whose stability is studied. This evolution is generated by initial conditions incorporating the effect of the disturbances preceeding the considered initial moment.

Another field of modern engineering in which the stability problem holds a central place is *electric power engineering.* Actually, an electric energy system is an interconnection of electric generators with various prime movers (steam or water turbines), supplying with electric energy of constant voltage and frequency the numerous and various industrial or private users.

The components of the electric energy system can be described by various technical parameters, among which only the voltage and the frequency have to be kept constant, the other ones can be manipulated in order to achieve this goal. But the operating regimes of any electric energy system are always subject to disturbances such as switching off or switching in the users, short–circuits and other faults.

The requirement of keeping some parameters (as the frequency and the voltage) constant under acting disturbances clearly leads to a stability problem. This problem was studied from different points of view; from here quite a large amount of stability concepts used in electric power engineering emerged. Here we would like to point out that these concepts can be deduced from a single one.

The classical work on the stability of power systems (e.g., S.B. Crary, 1947) starts with an attempt to classify the disturbances of the power systems. A distinction is made between slow and fast load changes.

The disturbances generated by slow load changes are associated with so–called steady–state stability, while the fast load changes are associated with dynamic stability (sometimes also called transient stability). In other studies the two concepts are associated with the amplitude of the disturbances: static stability corresponds to small amplitude and dynamic stability corresponds to large amplitude disturbances.

In fact, as it was pointed out in electric energy engineering, there is a unique stability problem associated to the transition from one operating regime to another one. Due to the complexity of any electric energy system (hundreds of generators and thousands of users), in stability studies simplified models are introduced. The most natural simplified model is to consider a single synchronous generator connected either to another generator of comparable power (when the power of the equivalent network is comparable to the power of the considered generator) or to an infinite bus (when the power of the equivalent network is considerably larger than the power of the generator). Especially this last concept was proven to be very fruitful in stability studies of electric energy systems, because it allowed us to apply all known results from the analysis of a single electrical machine.

As it was mentioned before, the electric energy system is subject to various disturbances, both electrical (short–circuits, switching in or switching off various users) and mechanical (non–constant active torque, transitions of the active torque, speed governor compensating signals) which generate transients. The transients can result either in a new normal steady–state operation or in an abnormal steady–state operation, leading finally to frequency instability.

Among the transitions from one steady–state regime to another a distinction is made between simple transitions and complex transitions (A.A.Gorev, 1950).

By a *simple transition* one understands the transient process from one steady state to another under persistent disturbances. This is the case, for instance, of the switching of some network elements without short circuits or reclosure. The disturbance is not too large and the new steady state is quite close to the initial one. It is thus possible to make use of the *steady–state stability* concept.

A *complex transition* has at least two phases: the fault and the post–fault recovery (e.g., a short–circuit and the post–fault pendulation). In this case not only the disturbances are large but even the new steady–state can be very far from the initial one. In this situation the concept of *transient stability* is used.

However, the concept of transition clearly shows that in both cases the same problem is formulated: the stability problem. Indeed, in both cases an initial and a final steady–state are considered regardless of their being close (in some sense). If the disturbances are short–period, the final steady state coincides with the initial one. Under the action of the disturbances the system reaches some state representing the initial conditions for the transient process to the new steady state. Therefore in both cases one takes into account the transitions in a neighbourhood of a steady state generated by disturbances whose effect is incorporated in the initial conditions.

A joint field of *electric and electronic engineering* where a stability problem also occurs is *circuit theory.* A circuit is a connection of resistors, capacitors, inductors, transistors, voltage and current sources, etc. having definite properties. The stability problem of circuit theory results from some practical aspects of circuit analysis.

Consider, for instance, the computation of the steady periodic solution of a nonlinear circuit with periodic (eventually harmonic) sources. The initial conditions corresponding to this solution are obviously unknown, hence arbitrary initial conditions are chosen and the differential equations of the nonlinear circuit are integrated on a digital computer until the solution is sufficiently close to the periodic solution.

This approach seems quite efficient, but in fact its success depends on the positive answer to the following problems:

1. Existence of solutions for the differential equations of the circuit with arbitrary initial conditions. This can be considered also a first test for the mathematical model. The circuits with this property are called definite.

2. Uniqueness of the steady periodic solution and asymptotic convergence to it of all other solutions of the system.

This property is called convergence.

Starting from the convergence property one can introduce the stability problem of circuit theory. Indeed, if the steady periodic solution is considered to be the basic regime of the circuit, any evolution determined by some initial conditions represents a disturbed evolution (motion) with respect to the basic regime. The property that any disturbed motion tends to the basic one is very similar to the properties previously seen in other engineering systems.

What is different is that here we consider a steady periodic motion, while in the previous cases a constant steady state was considered to be the basic regime. But electric circuits also have constant steady states: this occurs for circuits with constant sources. In both cases the stability problem can be introduced as before: a steady state motion* under short–period disturbances is considered.

Under the action of disturbances the system reaches some state which can be considered the initial state when the disturbance vanishes. This initial condition generates a disturbed motion with respect to the steady state motion and again we have to study in a neighbourhood of the basic motion the evolution of a system disturbed by initial conditions.

The engineering field in which stability is one of the main problems is *automatic control*. The above examined engineering systems – electrical, hydraulic, nuclear, chemical – require keeping fixed some parameters at given values, independently of the acting disturbances (e.g., the nuclear reactor power level, the frequency of the electric energy system).

The operation of industrial plants also requires manoeuvres – programmed modifications of some technical parameters. These modifications, as well as keeping fixed other parameters at given values, are not performed manually but automatically, using feedback controllers.

---

*From the point of view of dynamical systems theory both types of steady motion belong to the recurrent motions. Worthy of mention is that another type of recurrent motion – almost periodic motion – can be seen in circuit theory: the steady motion of circuits with modulated sources.

These controllers are working based on the so–called *error control principle*, sometimes also called the principle of Watt. According to this principle, the parameter that must be kept constant is permanently compared to a reference signal and the resulting error is amplified in order to be able to drive the plant actuators. The control is performed in order to make the error vanish, regardless the disturbances that act on the plant. Therefore a feedback structure is obtained.

It is obvious that by modifying the reference signal such a structure can also be used for manoeuvres. This is the case in practice when realizing the control of large power plants using reduced power level control signals. It has been shown that in the case of nuclear reactors a feedback structure can lead to instability. This holds for control systems too. If the controller gain is quite large – imposed by the required output power level of this block – the small variations of the controlled parameter will be amplified and, being applied to the plant input, will generate new variations which, instead of attenuating the controlled parameter variations, will amplify them. This phenomenon, called regeneration (H. Nyquist, 1932), shows an ambiguous role of the feedback: it reduces the effect of exogenous disturbances but it can destabilize a system composed of stable components. On the other hand, feedback control systems are introduced also for improving plant stability. This is possible by suitably choosing the controller structure and parameters. Therefore feedback structures are often met in stabilization problems.

The classical point of view – originating from mechanics – is valid for control systems too. Considering that the control device has to preserve some motion of the system (or some steady–state), this basic motion is subject to disturbances. If it is possible to consider these disturbances as being short–period, then their effect can be incorporated, as in the previous cases, in the initial conditions.

It was shown that a feedback structure is present both in nuclear reactors and in control systems. The difference between this two cases is that the controller block of the feedback structure can be chosen by the designer in order to obtain desired performance and stability for the plant, while the feedback structure of the nuclear reactor is an internal one. This last case corresponds to classical dynamical systems which are in some sense "self–closed", subject to their own, internal laws and their initial conditions.

The importance of the stability problem in automatic control – due to specific practical problems (control and stabilization) as well as to specific (feedback) structure – generated a new understanding of the stability concept even in those fields in which this concept was considered in the best case as secondary. The discovery of internal feedback structures in various physical systems made feedback an important concept for explaining the stability–instability mechanism of these systems (Yu.I. Neymark, 1978).

References

ARIS, R. (1961) *The Optimal design of Chemical Reactors*, Acad. Press, N.Y.

ARONOVICH, G. V., KARTVELISHVILI, N. A., LYUBIMTSEV, G. A. (1968) *Hydraulic shock and surge tanks* (in Russian), Nauka, Moscow.

CRARY, S.B. (1947) *Power System Stability*, J. Wiley and Sons, N.Y.

GOREV, A.A. (1950) *Transient processes of the synchronous machine* (in Russian), Gosenergoizdat, Moscow.

MALKIN, I.G. (1952) *Stability of motion* (in Russian), Gostiehizdat, Moscow.

NEYMARK, YU.I. (1978) *Dynamic systems and controlled processes* (in Russian), Nauka, Moscow.

NYQUIST, H.J. (1932) *Regeneration Theory*, BSTJ vol.11, pp. 126–147.

CHAPTER 2

Some General Results in Stability Theory

2.1 Basic Concepts

Consider a system

$$y' = g(t, y) \tag{2.1.1}$$

with $g : R_+ \times R^n \longrightarrow R^n$ continuous. Let $\hat{y}$ be a solution defined on $[t_0, \infty)$.

Definition 2.1 *The solution $\hat{y}$ of (2.1.1) is called* <u>*Liapunov stable*</u> *if for each $\epsilon > 0$ there exists $\delta(\epsilon, t_0)$ such that if*

$$|y_0 - \hat{y}(t_0)| < \delta(\epsilon, t_0) ,$$

then

$$|y(t;\ t_0, y_0) - \hat{y}(t)| < \epsilon \qquad \text{for all } t \geq t_0;$$

here $y(\cdot;\ t_0, y_0)$ is the solution of (2.1.1) such that $y(t_0;\ t_0, y_0) = y_0$.

If δ does not depend on t_0, the stability is *uniform.*

Note that the above definition corresponds to the requirements deduced from the analysis of the stability problems discussed in Chapter 1:

the effect of short–period disturbances and manoeuvres was incorporated in the initial conditions, and stability means small deviation from the basic motions for sufficiently small deviations of the initial conditions.

The general concept of stability of a solution can be reduced to the one of a special equilibrium, corresponding to a solution which identically equals zero. Indeed, with (2.1) we can associate a new system

$$x' = f(t,x) \tag{2.1.2}$$

where $f(t,x) = g(t, x + \hat{y}(t)) - g(t, \hat{y}(t))$.

We see that $f(t,0) \equiv 0$, hence the function defined by $x(t) \equiv 0$ is a solution of (2.1.2). We see next that if $y(t)$ is a solution to (2.1.1), then $x(t)$ defined by $x(t) = y(t) - \hat{y}(t)$ is a solution of (2.1.2). Conversely, if $x(t)$ is a solution of (2.1.2), then $y(t)$ defined by $y(t) = x(t) + \hat{y}(t)$ is a solution of (2.1.1). We see that stability of the solution $\hat{y}$ of (2.1.1) is equivalent to the stability of the solution $x \equiv 0$ of (2.1.2). This is why in the future, without loss of generality, we can consider only the stability properties of the solution which is identically equal to zero.

Definition 2.2 *The solution $x \equiv 0$ of (2.1.2) is asymptotically stable if it is stable and, moreover, there exists $\delta_0 > 0$ such that if $|x_0| < \delta_0$, then* $\lim_{t\to\infty} x(t; t_0, x_0) = 0$

An important concept is uniform asymptotic stability.

Definition 2.3 *The solution $x \equiv 0$ to (2.1.2) is uniformly asymptotically stable if there exist $\delta_0 > 0$ and functions $\delta(\cdot)$, $T(\cdot)$ such that $|x_0| < \delta(\epsilon)$ implies $|x(t; t_0, x_0)| < \epsilon$ for all $t \geq t_0$ and if $|x_0| < \delta_0$ then $|x(t; t_0, x_0)| < \epsilon$ for $t \geq t_0 + T(\epsilon)$.*

We note that uniform asymptotic stability means uniform Liapunov stability and an asymptotic behaviour which is independent both of the initial moment and of the initial conditions of the perturbed motion provided these initial conditions are small enough.

We will finally consider the strongest stability property which corresponds not only to uniform asymptotic stability but gives also quantitative description of the behaviour of solutions.

Definition 2.4 *The solution* $x \equiv 0$ *of* (2.1.2) *is exponentially stable if there exist* $\delta_0 > 0$, $\alpha_0 > 0$, $\beta \geq 1$ *such that for all* x_0 *with* $|x_0| < \delta_0$ *we have* $x(t; t_0, x_0)| \leq \beta e^{-\alpha(t-t_0)}|x_0|$ *for* $t \geq t_0$.

2.2 Linear Systems with Constant Coefficients. Stability by the First Approximation.

Consider the special case of a system (2.1.2) for which f does not depend on the first argument. Such system is sometimes called autonomous. If the system admits the solution $\hat{x} \equiv 0$ and if we assume that f is C^1 then we may write

$$f(x) = Ax + F(x), \qquad A = \frac{\partial f}{\partial x}(0)$$

Since the matrix A is defined to be the Jacobian matrix, F has the property that for every $\gamma > 0$ there exists $\delta(\gamma) > 0$ such that if $|x| < \delta(\gamma)$ then $|F(x)| \leq \gamma|x|$. This last property shows that F corresponds to "higher order terms" which are small with respect to the linear ones in the neighbourhood of the equilibrium $x \equiv 0$, hence it is natural to start by considering the "first approximation" corresponding to the linear system

$$x' = Ax \tag{2.2.1}$$

For such systems we have a full description of the solutions in terms of the spectral properties of A. The most important fact in this context is that for (2.2.1), exponential stability is equivalent to the property that $\Re\lambda_j \leq -\alpha < 0$ for all eigenvalues λ_j of matrix A. We will not dwell upon this very well-known fact; for detailed discussion the reader can consult a number of classical monographs in stability theory (Četaev,

1946; Bellman, 1953; Malkin, 1952; Halanay, 1966; Hahn, 1967 and many others); we will recall only that the property mentioned above means that all roots of the characteristic equation of A,

$$\det(\lambda I - A) = 0$$

have strictly negative real parts.

In order to check this property one may use the general *Hurwitz condition*: Consider the polynomial

$$p(z) = a_0 z^n + \ldots + a_n$$

with *real* coefficients and $a_0 > 0$. With this polynomial we associate the Hurwitz determinants

$$D_1 = a_1, \qquad D_2 = \begin{vmatrix} a_1 & a_0 \\ a_3 & a_2 \end{vmatrix}, \qquad D_3 = \begin{vmatrix} a_1 & a_0 & 0 \\ a_3 & a_2 & a_1 \\ a_5 & a_4 & a_3 \end{vmatrix}, \ldots$$

$$D_n = \begin{vmatrix} a_1 & a_0 & 0 & \ldots & 0 \\ a_3 & a_2 & a_1 & \ldots & 0 \\ \vdots & \vdots & \vdots & \ldots & \vdots \\ a_{2n-1} & a_{2n-2} & a_{2n-3} & \ldots & a_n \end{vmatrix}, \qquad a_k = 0,\ k > n$$

Then a necessary and sufficient condition for all roots of the polynomial to lie in the half plane $\Re z < 0$ is that $D_k > 0$ for all $k \leq n$.

There exist several proofs of this result which can be also found in various monographs. We will not reproduce here any of these proofs because we are not going to use this result here; we have stated this condition just in order to show how complicated it may be to check effectively the stability of a higher order system by using Hurwitz determinants.

Let us state and prove now the main result which shows when we are allowed to reduce the study of stability to the case of a linear system with constant coefficients.

Theorem 2.1 *Consider a system*

$$x' = Ax + F(x) \tag{2.2.2}$$

where A *is a* $n \times n$ *matrix and* F *is continuous and has the property that for every* $\gamma > 0$ *there exists* $\delta(\gamma) > 0$ *such that if* $|x| < \delta(\gamma)$ *then* $|F(x)| \le \gamma |x|$.

Assume that all roots λ_j *of the polynomial* $\det(\lambda I - A)$ *satisfy* $\Re\lambda_j \le -2\alpha < 0$.

Then there exist $\delta_0 > 0$, $\beta \ge 1$ *such that if* $|x_0| < \delta_0$, *then it follows that*

$$|x(t; t_0, x_0)| \le \beta e^{-\frac{\alpha}{2}(t-t_0)} |x_0|, \qquad t \ge t_0$$

Proof Let $x(\cdot; t_0, x_0)$ be a solution of (2.2.2). We can write

$$\frac{d}{dt} x(t; t_0, x_0) = Ax(t; t_0, x_0) + h(t),$$

$$h(t) = F(x(t; t_0, x_0)),$$

hence, by using the variations of constants formula

$$x(t; t_0, x_0) = e^{A(t-t_0)} x_0 + \int_{t_0}^{t} e^{A(t-s)} h(s) ds,$$

that is:

$$x(t; t_0, x_0) = e^{A(t-t_0)} x_0 + \int_{t_0}^{t} e^{A(t-s)} F(x(s; t_0, x_0)) ds.$$

The assumption concerning the roots of the characteristic polynomial implies that there exists $\beta \ge 1$ such that $\left| e^{A(t-s)} \right| \le \beta e^{-\alpha(t-s)}$ for all $t \ge s$. We deduce that

$$|x(t; t_0, x_0)| \le \beta e^{-\alpha(t-t_0)} |x_0| + \int_{t_0}^{t} \beta e^{-\alpha(t-s)} |F(x(s; t_0, x_0))| ds.$$

We take $\gamma = \frac{\alpha}{2\beta}$, $\delta_0 = \frac{\delta(\gamma)}{2\beta}$ and assume that $|x_0| < \delta_0$; then there exists a maximal interval $[t_0, t_0 + T)$ such that $|x(t; t_0, x_0)| < \delta(\gamma)$ for all $t \in [t_0, t_0 + T)$. For $t \in [t_0, t_0 + T)$ we thus have $|F(x(s; t_0, x_0))| \leq \gamma |x(s; t_0, x_0)|$, $t_0 \leq s \leq t$, hence,

$$|x(t; t_0, x_0)| \leq \beta e^{-\alpha(t-t_0)}|x_0| + \beta\gamma \int_{t_0}^{t} e^{-\alpha(t-s)}|x(s; t_0, x_0)|ds$$

or

$$e^{\alpha t}|x(t; t_0, x_0)| \leq \beta e^{\alpha t_0}|x_0| + \beta\gamma \int_{t_0}^{t} e^{\alpha s}|x(s; t_0, x_0)|ds.$$

The Gronwall - Bellman lemma leads to

$$e^{\alpha t}|x(t; t_0, x_0)| \leq \beta e^{\alpha t_0}|x_0| e^{\beta\gamma(t-t_0)}$$

that is

$$|x(t; t_0, x_0)| \leq \beta e^{-(\alpha-\beta\gamma)(t-t_0)}|x_0| < \beta e^{-\frac{\alpha}{2}(t-t_0)}|x_0| \tag{2.2.3}$$

(since $\beta\gamma < \frac{\alpha}{2}$).

The above inequality holds for $t_0 \leq t < t_0 + T$: if T were finite then we would have

$$|x(t_0 + T; t_0, x_0)| = \delta(\gamma)$$

But from (2.2.3) we deduce

$$|x(t_0 + T; t_0, x_0)| \leq \beta e^{-\frac{\alpha}{2}T}|x_0| \leq \beta\delta_0 = \frac{\delta(\gamma)}{2},$$

which is a contradiction. It follows that $T = \infty$ and

$$|x(t; t_0, x_0)| \leq \beta e^{-\frac{\alpha}{2}(t-t_0)}|x_0| \quad \text{for } t \geq t_0$$

as stated.

Now, we will discuss a problem in engineering where this general mathematical result is applied. Consider a control system

$$x' = f(x, u) \tag{2.2.4}$$

where u is a control parameter. We will not discuss the different interpretations this control parameter may have in specific situations. We note only that for a fixed value $\hat{u}$ of this parameter, the constant solutions, corresponding to a steady–state regime are given by the system

$$f(x, \hat{u}) = 0 \tag{2.2.5}$$

Let us assume that this system has a solution $\hat{x}$; then, if we fix the control parameter $\hat{u}$, the system (2.2.4) will have the constant solution $\hat{x}$ and if this constant solution is such that the eigenvalues of the Jacobian matrix $\frac{\partial f}{\partial x}(\hat{x}, \hat{u})$ have strictly negative real parts, by Theorem 2.1 we deduce that $\hat{x}$ is exponentially stable, that is, for small enough perturbations the evolution will not take place far away from $\hat{x}$ and moreover, the perturbations will decay exponentially.

In the engineering practice the situation often occurs when we want to pass from a given steady state operation to another one; to this end we replace the control value $\hat{u}$ by a new value $\tilde{u}$ such that the equation (2.2.5) has the desired solution $\tilde{x}$, corresponding to the desired steady–state regime. We assume even that $\frac{\partial f}{\partial x}(\tilde{x}, \tilde{u})$ has also all eigenvalues with strictly negative real parts, that is $\tilde{x}$ is exponentially stable. Can one be sure that by passing directly from the value $\hat{u}$ of the control to the value $\tilde{u}$, the evolution will be "directed to" $\tilde{x}$?

In fact, when the control is set at $\tilde{u}$ the former steady state $\hat{x}$ becomes an initial condition for the evolution given by (2.2.4) for $u = \tilde{u}$; even if the condition for the first approximation is satisfied one knows only that the evolution will lead to $\tilde{x}$ if the initial condition is close enough to $\tilde{x}$, hence the change of regime will be ensured only if $\hat{x}$ is close enough to $\tilde{x}$. It is why one cannot expect that such sudden manoeuvre from the steady state (operation point) $\hat{x}$ to the new one $\tilde{x}$ will be successful. Tus, it may be of interest to see that under some natural conditions a *gradual manoeuvre* may be suitable.

Consider a C^2 function $f : X \times U \longrightarrow X$ with the following properties:

a) there exists a continuous branch $\hat{x}(\cdot)$ of solutions of (2.2.5) such that $\hat{x}(u)$ is continuous and

$$f(\hat{x}(u), u) = 0, \quad u \in \hat{U} \subset U$$

where $\hat{U}$ is compact;

b) the Jacobian matrix $\frac{\partial f}{\partial x}(\hat{x}(u), u)$ has all eigenvalues in a half-plane $\Re z \leq -\alpha < 0$ for all $u \in \hat{U}$.

It is now possible to state

Theorem 2.2 *Under the above assumptions there exist* $\hat{\rho} > 0$, $\hat{\beta} > 0$, $\hat{\alpha} > 0$ *such that for all* $u \in \hat{U}$ *if*

$$|x_0 - \hat{x}(u)| \leq \hat{\rho}$$

then

$$|x(t; t_0, x_0; u) - \hat{x}(u)| \leq \hat{\beta} e^{-\hat{\alpha}(t-t_0)} |x_0 - \hat{x}(u)|$$

for all $t \geq t_0$.

Proof One can write

$$f(x + \hat{x}(u), u) = A(u)x + F(x, u),$$

$$A(u) = \frac{\partial f}{\partial x}(\hat{x}(u), u), \qquad |F(x, u)| \leq \hat{M}|x|^2,$$

for $|x| \leq \hat{\rho}$ (it was assumed f is C^2 and M is obtained from an estimate for the second derivatives).

From continuity and compactness of $\hat{U}$, we deduce that

$$\left|e^{A(u)t}\right| \leq \hat{\beta} e^{-2\hat{\alpha}t}, \quad t \geq 0, \quad \forall u \in \hat{U}$$

The conclusion follows from the explicit formulae in the proof of Theorem 2.1.

From Theorem 2.2 we deduce the existence of $\hat{\delta} > 0$ such that if $|\hat{u} - u| < \hat{\delta}$ then the manoeuvre from $\hat{x}(\hat{u})$ to $\hat{x}(\tilde{u})$ will work and viceversa.

Indeed, since $\hat{x}(\cdot)$ is continuous on the compact set $\hat{U}$, it is uniformly continuous, hence there exists $\hat{\delta}$ such that $|\hat{u} - u| \leq \hat{\delta}$ implies $|\hat{x}(\hat{u}) - \hat{x}(\tilde{u})| \leq \hat{\rho}$, where $\hat{\rho}$ is from Theorem 2.2. It is this inequality that ensures the desired behaviour of the evolution according to Theorem 2.1.

The practical meaning of this result is that if one wants to safely pass from $\hat{x}(\hat{u})$ to $\hat{x}(\tilde{u})$ then one has to work gradually, considering a sequence of intermediate manoeuvres with:

$$u_0 = \hat{u}, \ldots, u_N = \tilde{u}$$

and $|u_j - u_{j+1}| \leq \hat{\delta}$.

2.3 Liapunov Functions

We will present here some results concerning the so called method of Liapunov functions which will be of general use in the applications we want to describe.

The background of this method originates from Mechanics: if the energy of an isolated physical system is decreasing for any of its states except of equilibrium one, then it will decrease until it will reach its minimum – corresponding to the equilibrium state.

By generalizing this idea in a natural way, Liapunov introduced some state functions which are energy–like i.e. of constant sign and are decreasing along the trajectories of the system of differential equations. In spite of these similarities, these state functions are no longer connected with physics, in particular with energy.

Theorem 2.3 *Consider a system*

$$x' = f(x), \quad f : D \subset R^n \longrightarrow R^n, \quad f(0) = 0 \tag{2.3.1}$$

and assume there exists a C^1 *function* $V : D \longrightarrow R$ *with the following properties:*

a) $\alpha(|x|) \leq V(x) \leq \beta(|x|)$ *for all* $x \in D$, *where* α, β *are defined on* $[0, \infty)$, *continuous, strictly increasing and* $\alpha(0) = \beta(0) = 0$ *;*

b) $\dfrac{\partial V}{\partial x}(x) \cdot f(x) \leq 0$.

Then the solution $x \equiv 0$ *of* (2.3.1) *is uniformly stable in the sense of Definition* 2.1.

Proof Let $\epsilon > 0$ and choose $\delta(\epsilon) = \beta^{-1}(\alpha(\epsilon))$.

We assume that $|x_0| < \delta(\epsilon)$ and consider a solution $x(\cdot\,; t_0, x_0)$ of (2.3.1). Let $\hat{V}(t) = V(x(t; t_0, x_0))$; we have

$$\hat{V}'(t) = \frac{\partial V}{\partial x}(x(t; t_0, x_0))\frac{d}{dt}x(t; t_0, x_0) =$$

$$= \frac{\partial V}{\partial x}(x(t; t_0, x_0))f(x(t; t_0, x_0)) \leq 0.$$

Hence $t \longrightarrow \hat{V}(t)$ is decreasing. We deduce that $\hat{V}(t) \leq \hat{V}(t_0)$ for $t \geq t_0$. Therefore

$$\alpha(|x(t; t_0, x_0)|) \leq V(x(t; t_0, x_0)) = \hat{V}(t) \leq \hat{V}(t_0) =$$

$$= V(|x_0|) \leq \beta(|x_0|)$$

It follows that

$$\alpha(|x(t; t_0, x_0)|) \leq \beta(|x_0|) \leq \beta(\delta(\epsilon)) = \alpha(\epsilon)$$

hence $|x(t; t_0, x_0)| < \epsilon$ for $t \geq t_0$ and uniform stability is proved.

Theorem 2.4 *Under assumptions of Theorem* 2.3 *let there exist a continuous, increasing function* γ, *defined on* $[0, \infty)$, $\gamma(0) = 0$ *such that*

$$\frac{\partial V}{\partial x}(x)f(x) \leq -\gamma(|x|)$$

Then the solution $x \equiv 0$ *of (*2.3.1*) is uniformly asymptotically stable in the sense of Definition* 2.3.

Proof Take $h > 0$ such that the set $\{x : |x| \le h\}$ is included in D. Define $\delta(\epsilon) = \min\left\{\beta^{-1}(\alpha(\epsilon)),\, \beta^{-1}(\alpha(h))\right\}$, $\delta_0 = \beta^{-1}(\alpha(h))$ and $T(\epsilon) = \frac{\beta(\delta_0)}{\gamma(\delta(\epsilon))} = \frac{\alpha(h)}{\gamma(\delta(\epsilon))}$. Let us show that δ_0 and T so defined have all properties required in Definition 2.3: $|x_0| < \delta_0$ and $t \ge t_0 + T(\epsilon)$ imply $|x(t; t_0, x_0)| < \epsilon$. We will prove existence of $\hat{t} \in [t_0, t_0 + T(\epsilon)]$ such that $|x(\hat{t}; t_0, x_0)| < \delta(\epsilon)$; from the definition of $\delta(\epsilon)$ it will follow that

$$|x(t; t_0, x_0)| = |x(t; \hat{t}, x(\hat{t}; t_0, x_0))| < \epsilon$$

for $t > \hat{t}$ hence also $|x(t; t_0, x_0)| < \epsilon$ for $t \ge t_0 + T(\epsilon)$.

Assume that such $\hat{t}$ does not exist, that is

$$|x(t; t_0, x_0)| \ge \delta(\epsilon) \qquad \text{for all} \qquad t \in [t_0, t_0 + T(\epsilon)]\,.$$

With the same notations as in the proof of the Theorem 2.3, we will have

$$\hat{V}'(t) = \frac{\partial V}{\partial x}(x(t; t_0, x_0))f(x(t; t_0, x_0)) \le -\gamma(|x(t; t_0, x_0)|)$$

Since $|x(t; t_0, x_0)| \ge \delta(\epsilon)$ it follows that $\gamma(|x(t; t_0, x_0)|) \ge \gamma(\delta(\epsilon)) = \frac{\beta(\delta_0)}{T(\epsilon)}$. Hence

$$\hat{V}'(t) \le -\frac{\beta(\delta_0)}{T(\epsilon)}, \quad \hat{V}(t_0 + T(\epsilon)) - \hat{V}(t_0) \le -\beta(\delta_0),$$

$$\hat{V}(t_0 + T(\epsilon)) \le \hat{V}(t_0) - \beta(\delta_0) =$$

$$= V(x_0) - \beta(\delta_0) \le \beta(|x_0|) - \beta(\delta_0) \le 0,$$

in contradiction with

$$\hat{V}(t_0 + T(\epsilon)) = V(x(t_0 + T(\epsilon); t_0, x_0)) \ge$$

$$\ge \alpha(|x(t_0 + T(\epsilon); t_0, x_0)|) > 0$$

This ends the proof.

In many applications it is important to have *global asymptotic stability*, that is δ_0 may be taken arbitrary. Such property holds if

$\lim_{r\to\infty}\alpha(r) = \infty$; indeed, in such case we have also $\lim_{r\to\infty}\beta(r) = \infty$, $\beta : [0,\infty) \longrightarrow [0,\infty)$ and $\lim_{h\to\infty}\beta^{-1}(\alpha(h)) = \infty$; if h may be taken arbitrarily large (that is $f : R^n \longrightarrow R^n$) then δ_0 may be taken arbitrarily large.

In many applications, Liapunov functions associated in a natural way to specific models do not satisfy all requirements in Theorem 2.4. It thus appears useful to weaken these requirements while preserving the conclusion concerning asymptotic stability. Such results were first obtained by Barbašin and Krasovskii (1952) and have been further developed by La Salle (1968).

In what follows, we will call a function $V : D \subset R^n \longrightarrow R$ a Liapunov function for (2.3.1) if it is C^1, $\frac{\partial V}{\partial x}(x)f(x) \leq 0$ for all $x \in D$, and if for every $\hat{x} \in D$ there exists a neighbourhood $\mathcal{U}$ of $\hat{x}$ and $\mu \in R$ such that $V(x) \geq \mu$ for all $x \in \mathcal{U}$.

Let

$$G = \left\{x \in D, \frac{\partial V}{\partial x}(x)f(x) = 0\right\}.$$

Theorem 2.5 *Let* V *be a Liapunov function for* (2.3.1), G *defined as above. Then every solution of* (2.3.1) *which is defined on an interval* (t', t'') *and such that for* $t \in (t_0, t'')$ $x(t) \in D$ *will have the following asymptotic behaviour:*

a) either $\lim_{t\to t''}|x(t)| = \infty$, *or*

b) $t'' = \infty$ *and* $\lim_{t\to\infty} d(x(t), M \cup \{\infty\}) = 0$ *where* M *is the largest invariant set (with respect to the flow defined by* (2.3.1)*) contained in* G.

Proof Let us recall that M is invariant with respect to the flow defined by (2.3.1) if it follows from $x(t_0) \in M$ that $x(t) \in M$ for all t for which the solution is defined. A set Ω will be called ω–limit for a solution $x(\cdot)$ if for each $\hat{x} \in \Omega$ there exists a sequence $\{t_k\}_k$, $\lim_{k\to\infty} t_k = t''$ such that $\lim_{k\to\infty} x(t_k) = \hat{x}$, and if it is the maximal set with the required property.

Notice that if the ω–limit set Ω is empty then $\lim_{t\to t''}|x(t)| = \infty$; indeed if this property did not hold, a sequence $(t_k)_k$ with $t_k \longrightarrow t''$ could be obtained such that $(x(t_k))_k$ is bounded. This bounded sequence would have a subsequence converging to a point which, by definition would belong to Ω, a contradiction with the assumption that Ω is empty. It remains to prove that b) holds if Ω is not empty. Let $\hat{x} \in \Omega$ and, correspondingly, let $(t_k)_k$ be increasing, $\lim_{k\to\infty} t_k = t''$, $\lim_{k\to\infty} x(t_k) = \hat{x}$, that is $|x(t_k)-\hat{x}| < \epsilon/2$ for $k \geq K_\epsilon$. If there exists $\hat{t}$ such that $|x(t)-\hat{x}| < \epsilon/2$ for all $t > \hat{t}$ then by virtue of a general result, $t'' = \infty$, that is the solution is globally defined for $t > t'$. If such $\hat{t}$ does not exist, the values $x(t)$ will enter and leave the ball $\{x : |x-\hat{x}| < \epsilon/2\}$ indefinitely. Then, we deduce the existence of a new sequence $(\hat{t}_k)_k$ with $\hat{t}_k > t_k$ such that $|x(\hat{t}_k) - \hat{x}| > \epsilon$ for all k. Next, we deduce the existence of a third sequence $(\tilde{t}_k)_k$ with $\tilde{t}_k \in [t_k, \hat{t}_k]$ such that $|x(t) - \hat{x}| < \epsilon$ for $t \in [t_k, t_k)$ and $|x(\tilde{t}_k) - \hat{x}| = \epsilon$. By considering, if necessary, subsequences we have

$$\alpha)\ t_1 < \tilde{t}_1 < t_2 < \tilde{t}_2 \ldots < t_k < \tilde{t}_k < t_{k+1} < \ldots < t'' ,$$

$$\beta)\ |x(\tilde{t}_k) - x_k(t_k)| \geq \epsilon/2 ,$$

$$\gamma)\ |x(\tilde{t}_k) - x(t_k)| = \left|\int_{t_k}^{\tilde{t}_k} f(x(s))\,ds\right| \leq M_0(\tilde{t}_k - t_k) ,$$

where $M_0 = \sup\{|f(x)|, |x-\hat{x}| \leq \epsilon\}$.

From β) and γ) we deduce that $\epsilon/2 \leq M_0(\tilde{t}_k - t_k)$ hence $\tilde{t}_k - t_k \geq \epsilon/2M_0$, $\tilde{t}_k - t_1 \geq \frac{k\epsilon}{2M_0}$ and we must have $t'' = \infty$.

We have in fact proved that if $\Omega \neq \emptyset$ then the solution $x(\cdot)$ is globally defined on (t', ∞).

Let us now return to the sequence $(t_k)_k$ and note that since $|x(t_k) - \hat{x}| < \epsilon/2$ the existence of μ_ϵ such that $V(x(t_k)) \geq \mu_\epsilon$ follows. On the other hand

$$V(x(t_{k+1})) - V(x(t_k)) = \int_{t_k}^{t_{k+1}} \frac{\partial V}{\partial x}(x(t))\, f(x(t))\, dt \leq 0 ,$$

hence the sequence is decreasing and thus has a finite limit. But *the function* $t \longrightarrow V(x(t))$ is also decreasing hence it has either a finite limit or tends to $-\infty$; nevertheless, since for a sequence the limit is finite, it follows that $\lim_{t\to\infty} V(x(t))$ *is finite.*

We will prove now that $\Omega \subset G$; assume by contradiction that there exists $\hat{x} \in \Omega$ and $\hat{x} \notin G$. Since $\hat{x} \notin G$ we must have

$$\frac{\partial V}{\partial x}(\hat{x})\, f(\hat{x}) < 0$$

and there exists $\delta > 0$ and a neighbourhood of $\hat{x}$ such that

$$\frac{\partial V}{\partial x}(x)\, f(x) < -\delta$$

for x in this neighbourhood. We choose $\epsilon > 0$ such that the above inequality holds for $|x - \hat{x}| < \epsilon$. If there exists $\hat{t}$ such that we have $|x(t) - \hat{x}| < \epsilon/2$ for all $t > \hat{t}$ then

$$V(x(t)) - V(x(t_0)) = \int_{t_0}^{t} \frac{\partial V}{\partial x}(x(\tau))\, f(x(\tau))\, d\tau < -\delta(t - t_0),$$

$$V(x(t)) < V(x(t_0)) - \delta(t - t_0),$$

$\lim\limits_{t\to\infty} V(x(t)) = -\infty$, a contradiction.

If such $\hat{t}$ does not exist, we can consider again the sequences $\tilde{t}_k$ and t_k to deduce

$$V(x(\tilde{t}_k)) - V(x(t_k)) = \int_{t_k}^{\tilde{t}_k} \frac{\partial V}{\partial x}(x(t))\, f(x(t))\, dt < -\delta(\tilde{t}_k - t_k)$$

and since $\tilde{t}_k - t_k \geq \frac{\epsilon}{2M_0}$ it follows that

$$V(x(\tilde{t}_k)) - V(x(t_k)) \leq -\frac{\delta\epsilon}{2M_0}$$

which contradicts

$$\lim_{k\to\infty} V(x(\tilde{t}_k)) = \lim_{k\to\infty} V(x(t_k)) = \lim_{t\to\infty} V(x(t))$$

Thus, we have shown that existence of $\hat{x} \in G \setminus \Omega$ is leads to contradiction, hence $\Omega \subset G$; since Ω *is invariant* and M is the largest invariant set in G, we deduce that $\Omega \subset M$. It follows that either $\lim_{t\to\infty} |x(t)| = \infty$ or $\lim_{t\to\infty} d(x(t), \Omega) = 0$, that is $\lim_{t\to\infty} d(x(t), M) = 0$ which ends the proof.

Throughout the proof, the use was made of invariance properties of several sets among which Ω, the ω-limit set of a solution. For this reason Theorem 2.5 is called *the invariance principle.*

For applications, the following consequences of Theorem 2.5 are important.

Corollary 2.1 *If* D *is bounded, open and positively invariant, if* V *is a Liapunov function for* (2.3.1) *on* D, *and if* $M \subset D$, *then* M *is an attractor and* D *is contained in the basin of attraction of* M.

Proof Since $x(t_0) \in D$ implies $x(t) \in D$ for $t \geq t_0$, $x(\cdot)$ is bounded for $t \geq t_0$ (D is bounded) and from Theorem 2.5 it follows that $d(x(t), M) \to 0$ for $t \to \infty$ if $x(t_0) \in D$. Therefore M is an attractor and D lies in its basin of attraction which ends the proof.

In applications, D is chosen to be a connected component of a set of the form $\{x : V(x) < c\}$; such set will be (from the properties of V) open and invariant; to apply Corollary 2.1 one has only to check the boundedness of this set.

Corollary 2.2 *If* $M = \{\hat{x}\}$, $V(\hat{x}) = 0$, $V(x) > 0$ *for* x *in a neighbourhood of* $\hat{x}$, *then* $\hat{x}$ *is asymptotically stable. If, moreover,* $V(x) \geq \alpha(|x|)$ *with* α *increasing, continuous,* $\alpha(0) = 0$, *and* $\lim_{r\to\infty} \alpha(r) = \infty$ *then* $\hat{x}$ *is globally asymptotically stable.*

Proof We have Liapunov stability of $\hat{x}$ since $V(\hat{x}) = 0$, $V(x) > 0$ in a neighbourhood of $\hat{x}$ and $\frac{\partial V}{\partial x}(x)\, f(x) \leq 0$, for all x in this neighbourhood. That means that solutions starting in a neighborhood of $\hat{x}$ are bounded and from Theorem 2.5, $\lim_{t\to\infty} d(x(t), \hat{x}) = 0$, hence $\hat{x}$ is asymptotically stable.

An Application from Mechanics

Consider a mechanical system under the action of dissipative and gyroscopic forces. The Lagrange equations describing the motion are

$$\frac{d}{dt}\frac{\partial L}{\partial q'} - \frac{\partial L}{\partial q} + \frac{\partial \Lambda}{\partial q} + (q')^* R = 0 \tag{2.3.2}$$

where $L(q, q')$ is a C^2 function, $L(q, 0) = 0$, $\frac{\partial L}{\partial q'}(q, 0) = 0$, $\Lambda(q)$ is also C^2, $\Lambda(0) = 0$, $\frac{\partial \Lambda}{\partial q}(0) = 0$, $\frac{\partial^2 L}{\partial q' \partial q'}(q, q') \geq a_0 I$ (I denotes, as usually, the identity matrix), $a_0 > 0$, $\frac{\partial^2 \Lambda}{\partial q \partial q} \geq a_1 I$, $a_1 > 0$ and the matrix $R(q, q')$ satisfies the inequality

$$(q')^* R(q, q') q' \geq \alpha |q'|^2, \qquad \alpha > 0$$

The Lagrange equations are not in the normal Cauchy form; to obtain such form we use the Legendre transformation

$$p = \frac{\partial L}{\partial q'}(q, q')$$

Since $\frac{\partial L}{\partial q'}(q, 0) = 0$ and $\frac{\partial^2 L}{\partial q' \partial q'} \geq a_0 I$ we can use implicit function theorem to deduce existence of a function F of class C^1 with the properties:

$$F(q, 0) = 0, \qquad \frac{\partial L}{\partial q'}(q, F(q, p)) = p$$

We deduce that if (q, q') is a solution of the Lagrange equations and p is defined as above, we have

$$q'(t) = F(q(t), p(t))$$

$$p'(t) = \frac{d}{dt}\frac{\partial L}{\partial q'}(q(t), q'(t)) = \frac{\partial L}{\partial q}(q(t), q'(t)) -$$

$$-\frac{\partial \Lambda}{\partial q}(q(t)) - (q'(t))^* R(q(t), q'(t)) =$$

$$= \frac{\partial L}{\partial q}(q(t), F(q(t), p(t))) - \frac{\partial \Lambda}{\partial q}(q(t)) -$$

$$-F^*(q(t),p(t))R(q(t),F(q(t),p(t)))$$

We deduce that (q,p) is a solution of the system

$$q' = F(q,p)$$

$$p' = \frac{\partial L}{\partial q}(q,F(q,p)) - \frac{\partial \Lambda}{\partial q}(q) - F^*(q,p)R(q,F(q,p)) \qquad (2.3.3)$$

It is also clear that if (q,p) is a solution of system (2.3.3), then (q,q') solves the Lagrange equations (2.3.2). It appears also obvious from the above notations that we have adopted the convention that q,q' are column vectors, $\frac{\partial L}{\partial q}, \frac{\partial L}{\partial q'}, \frac{\partial \Lambda}{\partial q}$ are row vectors hence p is a row vector.

For system (2.3.3) we have the equilibrium corresponding to $q = 0$, $p = 0$ since $F(q,0) = 0$, $\frac{\partial L}{\partial q}(q,0) = 0$, $\frac{\partial \Lambda}{\partial q}(0) = 0$.

This equilibrium corresponds to $q = 0$, $q' = 0$.

Consider now the Liapunov function

$$V(q,p) = pF(q,p) - L(q,F(q,p)) + \Lambda(q) \qquad (2.3.4)$$

which is associated to the system in a natural way. We have

$$\frac{\partial V}{\partial q}(q,p)F(q,p) + \left[\frac{\partial L}{\partial q}(q,F(q,p)) - \frac{\partial \Lambda}{\partial q}(q) -\right.$$

$$\left. -F^*(q,p)R(q,F(q,p))\right]\frac{\partial V}{\partial p} = \left[p\frac{\partial F}{\partial q}(q,p) - \frac{\partial L}{\partial q}(q,F(q,p)) -\right.$$

$$\left. -\frac{\partial L}{\partial q'}(q,F(q,p))\frac{\partial F}{\partial q}(q,p) + \frac{\partial \Lambda}{\partial q}(q)\right] F(q,p) + \left[\frac{\partial L}{\partial q}(q,F(q,p)) -\right.$$

$$\left. -\frac{\partial \Lambda}{\partial q}(q) - F^*(q,p)R(q,F(q,p))\right]\left[F(q,p) + \frac{\partial F^*}{\partial p}(q,p)p^* -\right.$$

$$\left. -\frac{\partial F^*}{\partial p}(q,p)\left(\frac{\partial L}{\partial q'}(q,F(q,p))\right)^*\right] =$$

$$= \left[p\frac{\partial F}{\partial q}(q,p) - \frac{\partial L}{\partial q}(q,F(q,p)) - p\frac{\partial F}{\partial q}(q,p) + \frac{\partial \Lambda}{\partial q}(q)\right] F(q,p)+$$

$$+ \left[\frac{\partial L}{\partial q}(q,F(q,p)) - \frac{\partial \Lambda}{\partial q}(q) - F^*(q,p)R(q,F(q,p))\right]\left[F(q,p)+\right.$$

$$\left. + \frac{\partial F^*}{\partial p}(q,p)p^* - \frac{\partial F^*}{\partial p}(q,p)p^*\right] =$$

$$= -F^*(q,p)R(q,F(q,p))F(q,p) \leq -\alpha|F(q,p)|^2$$

In the above computation, the use was made of the identity

$$\frac{\partial L}{\partial q'}(q,F(q,p)) \equiv p$$

which defines $F(q,p)$.

The inequality obtained above shows that $V(q,p)$ is indeed a Liapunov function. It is then obvious that the derivative function of V (the left hand side of the above inequality) vanishes if and only if $F(q,p) = 0$; we have to look for the maximal invariant set with respect to the solutions of (2.3.3) satisfying this condition.

If $F(q(t),p(t)) \equiv 0$ then $q'(t) \equiv 0$ hence q is constant. On the other hand

$$p(t) = \frac{\partial L}{\partial q'}(q(t),F(q(t),p(t))) = \frac{\partial L}{\partial q'}(q(t),0) \equiv 0$$

hence the invariant set consists only of equilibrium points of the form $(q,0)$. But if p is constant and $F(q,p) = 0$ we have $\frac{\partial \Lambda}{\partial q}(q) = 0$. We write

$$\Lambda(q) = \left(\int_0^1 \frac{\partial \Lambda}{\partial q}(sq)ds\right) q$$

$$\frac{\partial \Lambda}{\partial q}(q) = \int_0^1 q^* \frac{\partial^2 \Lambda}{\partial q \partial q}(sq)ds$$

If $\frac{\partial \Lambda}{\partial q}(q) = 0$ we have also $\frac{\partial \Lambda}{\partial q}(q)q = 0$. Therefore

$$\int_0^1 q^* \frac{\partial^2 \Lambda}{\partial q \partial q}(sq)q ds = 0$$

But we assumed that $\frac{\partial^2 \Lambda}{\partial q \partial q} \geq a_1 I$ hence $q^* \frac{\partial^2 \Lambda}{\partial q \partial q}(sq)q \geq a_1 q^* q$ and $\frac{\partial \Lambda}{\partial q}(q) = 0$ implies $q = 0$ (at least locally). We deduce that the maximal invariant set for which $F(q,p) = 0$ consists of the equilibrium $(0,0)$.

We have further:

$$\Lambda(q) = \int_0^1 \int_0^1 s q^* \frac{\partial^2 \Lambda}{\partial q \partial q}(rsq) q dr ds \geq$$

$$\geq \int_0^1 \int_0^1 a_1 s |q|^2 dr ds = \frac{1}{2} a_1 |q|^2.$$

Denoting

$$\hat{V}(q,p) = pF(q,p) - L(q, F(q,p)) =$$

$$= \frac{\partial L}{\partial q'}(q, F(q,p))F(q,p) - L(q, F(q,p))$$

we have $V(q,p) = \Lambda(q) + \hat{V}(q,p)$.
On the other hand

$$L(q, F(q,p)) = L(q,0) + \frac{\partial L}{\partial q'}(q,0)F(q,p) +$$

$$+ \frac{1}{2} F^*(q,p) \left(\int_0^1 \frac{\partial^2 L}{\partial q' \partial q'}(q, sF(q,p)) ds \right) F(q,p)$$

and

$$\frac{\partial L}{\partial q'}(q, F(q,p))F(q,p) =$$

$$= \left(\int_0^1 F^*(q,p) \frac{\partial^2 L}{\partial q' \partial q'}(q, sF(q,p)) ds \right) F(q,p).$$

Taking into account the inequality satisfied by the matrix $\frac{\partial^2 L}{\partial q' \partial q'}$ and the fact that $L(q,0) = 0$, $\frac{\partial L}{\partial q'}(q,0) = 0$, we find

$$\hat{V}(q,p) \geq a_0 |F(q,p)|^2 - \frac{a_0}{2}|F(q,p)|^2 = \frac{a_0}{2}|F(q,p)|^2$$

hence

$$V(q,p) \geq a_1 |q|^2 + \frac{a_0}{2}|F(q,p)|^2$$

In fact, since $p = \frac{\partial L}{\partial q'}(q, F(q,p))$ we see that $F(q,p) = 0$ implies $p = 0$ and

$$V(q,p) \geq \alpha_0(|q|^2 + |p|^2)$$

The conclusion is that the equilibrium is globally asymptotically stable.

Remarks About Estimates of Liapunov Function

In both Theorems 2.3 and 2.4 we assumed existence – at least locally – of some functions which are monotonically increasing, continuous and zero when the argument is zero. Such functions, sometimes called Massera functions, play a central role in proving stability theorems. For a positively definite function $V : D \subset R^n \longrightarrow R_+$, that is satisfying $V(x) \geq 0$ and $V(x) = 0$ if and only if $x = 0$, such functions do always exist, provided V is continuous. Indeed, if $0 < r_1 < r_2$, for $r \in [r_1, r_2]$ we can define

$$\alpha(r) = \min_{r \leq |y| \leq r_2} V(y), \qquad \beta(r) = \sup_{|y| \leq r} V(y)$$

and $\alpha(r)$ and $\beta(r)$ are obviously monotonic. From the continuity of V one can obtain

$$\alpha(r) \leq V(x) \leq \beta(r), \qquad |x| = r$$

and, therefore

$$\alpha(|x|) \leq V(x) \leq \beta(|x|)$$

2.4 Application of Liapunov Functions in Some Problems of Hydraulic Engineering

1. It is a classical problem in Hydraulic Engineering to establish a critical area of the horizontal section of a surge tank in order to ensure stability when the hydropower station is regulated for constant power (Ch. Jaeger, 1977). The simplest situation corresponds to a surge tank and a tunnel and is described by a second order system of differential equations:

$$\frac{L}{g}\frac{dW}{dt} + Z + P'W^2 + R'\frac{dZ}{dt}\left|\frac{dZ}{dt}\right| = 0$$

$$F\frac{dZ}{dt} + \frac{N}{\eta g(H+Z)} - fW = 0\,. \tag{2.4.1}$$

In equations above, we made use of the notations common in Hydraulic Engineering: L is the length of the tunnel, f the area of the section in the tunnel, F the horizontal section of the surge tank, Z the level of the water in the basin, H is the height, N the power at the station after the manoeuvre, P', R' are loss coefficients, η is an efficiency coefficient, g is the acceleration of gravity; W is the speed of the water in the tunnel.

The equilibrium is obtained from the equations

$$\frac{N}{\eta g(H+Z)} = fW\,, \qquad Z + P'W^2 = 0. \tag{2.4.2}$$

We find

$$\frac{N^2}{\eta^2 g^2 f^2 (H+Z)^2} + \frac{Z}{P'} = 0\,,$$

hence

$$Z(H+Z)^2 + \frac{P'N^2}{\eta^2 g^2 f^2} = 0. \tag{2.4.3}$$

A simple discussion shows that if

$$\frac{4H^3}{27} > \frac{P'N^2}{\eta^2 g^2 f^2}$$

then (2.4.3) has a unique solution $Z_0 \in (-H/3, 0)$; the above condition is met in practice and such Z_0 corresponds to the equilibrium we want to study. Denoting

$$Y = Z - Z_0, \qquad V = \frac{dY}{dt},$$

the following second order system is obtained

$$\frac{dY}{dt} = V$$

$$F\frac{dV}{dt} = \frac{N}{\eta g (H_0 + Y)^2} V -$$

$$-\frac{fg}{L}\left[Y + Z_0 + \frac{P'}{f^2}\left(FV + \frac{N}{\eta g (H_0 + Y)}\right)^2 + R'V|V|\right], \tag{2.4.4}$$

where $H_0 = H + Z_0$.

With this system we associate the Liapunov function (A.Halanay and M.Popescu, 1979):

$$v(Y,V) = \frac{1}{2}FV^2 + \frac{gf}{2L}\left(1 - \frac{2P'N^2}{H_0^2 f^2 g^2 \eta^2 (H_0 + Y)}\right) Y^2 \tag{2.4.5}$$

If we compute

$$w(Y,V) = \frac{\partial v}{\partial Y}(Y,V)V + \frac{\partial v}{\partial V}(Y,V)\left\{\frac{N}{\eta g (H_0 + Y)^2} V -\right.$$

$$\left.-\frac{fg}{L}\left[Y + Z_0 + \frac{P'}{f^2}\left(FV + \frac{N}{\eta g (H_0 + Y)}\right)^2 + R'V|V|\right]\right\}\frac{1}{F}$$

we obtain

$$w(Y,V) = -\left[\frac{fgR'}{L}|V| + \frac{gP'F^2}{fL}V - \frac{N}{\eta g (H_0 + Y)^2} +\right.$$

$$\left.+\frac{2P'FN}{Lf\eta(H_0 + Y)}\right] V^2. \tag{2.4.6}$$

We restrict ourselves to the domain defined by

$$\frac{2P'N^2}{H_0^2 f^2 g^2 \eta^2 (H_0 + Y)} < 1$$

in order to have $\nu(Y, V) > 0$. Taking into account the definition of H_0 and the equation satisfied by Z_0, this condition reduces to

$$-\frac{2Z_0}{H_0 + Y} < 1$$

that is $Y > -H - 3Z_0$; note that since $Z_0 > -\frac{H}{3}$, we have $-3Z_0 - H < 0$ and our region is a strip containing the equilibrium $Y = 0$, $V = 0$.

In order to have $w(Y, V) < 0$ in a neighbourhood of the equilibrium, we must demand that

$$\frac{2P'FN}{Lf\eta H_0} > \frac{N}{\eta g H_0^2}$$

and we deduce the condition for the area F of the surge tank

$$F > \frac{1}{2g} \cdot \frac{fL}{P'H_0} = F_{Th} \tag{2.4.7}$$

where F_{Th} is called the section of Thoma which can be obtained from the simplest analysis of the linearized equations.

If we introduce the coefficient

$$\alpha = \frac{f^2 R'}{F^2 P'}$$

we can write

$$w(Y, V) = -\left[\frac{gP'F^2}{fL}(\alpha|V| + V) - \frac{N}{\eta g (H_0 + Y)^2} + \right.$$

$$\left. + \frac{2P'FN}{Lf\eta(H_0 + Y)^2}\right] V^2.$$

For $\alpha > 1$, the condition for w to be strictly negative is

$$\frac{2P'FN}{Lf\eta(H_0 + Y)} > \frac{N}{\eta g (H_0 + Y)^2}$$

and since $H_0 + Y > 0$ we have the simple condition

$$H_0 + Y > H_0 \frac{F_{Th}}{F}.$$

The case $\alpha > 1$ is not realistic; for $\alpha < 1$ we define the function

$$\Phi(Y, V) = V - \frac{fLN}{\eta g P' F^2 (1-\alpha)(H_0 + Y)} \left[\frac{1}{g(H_0 + Y)} - \frac{2P'F}{Lf} \right]$$

and if $\Phi(Y, V) > 0$, we have $w(Y, V) < 0$.

To discuss stability, including information concerning the domain of attraction we will use Theorem 2.5 (Corollary 2.1). Consider the family of curves (Ψ_c) defined by $v(Y, V) = c$, which are closed for c small enough, and then the domain in the interior of Ψ_c located in the domain defined by $\Phi(Y, V) > 0$. In this domain, the points where $w(Y, V) = 0$, correspond to $V = 0$, hence $Y = \text{const}$ (if we are on an invariant set) and from the equation for V we deduce that $Y = 0$, hence the attractor is the equilibrium.

Therefore, we proved asymptotic stability and obtained a procedure to estimate the basin of attraction.

2. In the early sixties, L.Escande and his school discussed more complicated schemes and among them the one of a surge tank fed by two tunnels; in this case the differential equation is of third order and the critical area was obtained by linearization and manipulation of the corresponding Hurwitz conditions (L.Escande et al, 1965). We will describe here the study of the general situation of a surge tank fed by an arbitrary number of tunnels; this study will show that a suitable Liapunov function makes it possible to obtain an estimate for the critical area from the linearized equations. This analysis exhibits some features related to the art of constructing Liapunov functions and we do not see any other procedure that could allow to obtain the result.

The mathematical model associated to the hydraulic scheme of a surge tank fed by n tunnels is

$$\frac{L_k}{g} \frac{dW_k}{dt} + P_k W_k^2 + Z = 0, \qquad k = 1, \ldots, n,$$

$$F\frac{dZ}{dt} + Q_T = \sum_1^n f_k W_k ,$$

$$\eta g Q_T(H + Z) = N \tag{2.4.8}$$

Here L_k is the length of the k-th tunnel, f_k the corresponding area of the transversal section; F is the area of the horizontal cross–section in the surge tank, W_k is the speed of the water in the k-th tunnel, Z is the water level in the surge tank, Q_T is the water flow at the engine (hydraulic turbine), N is the constant power, H the height of the surge tank, P_k are loss coefficients and η the efficiency coefficient.

The equilibria are defined by the equations

$$P_k W_k^2 + Z = 0 , \qquad Q_T = \sum_1^n f_k W_k ,$$

$$\eta g Q_T(H + Z) = N , \tag{2.4.9}$$

and we obtain the equation

$$Z(H + Z)^2 + \frac{N^2}{\eta^2 g^2 \left(\sum_1^n \frac{f_k}{\sqrt{P_k}}\right)^2} = 0 \tag{2.4.10}$$

If

$$\frac{N^2}{\eta^2 g^2 \left(\sum_1^n \frac{f_k}{\sqrt{P_k}}\right)^2} < \frac{4H^3}{27}$$

then the equation (2.4.10) has a root Z_0 in the interval $(-H/3, 0)$ corresponding to the equilibrium that has to be stable. After linearization around this equilibrium we obtain a linear system with constant coefficients of the form

$$\frac{L_k}{g}\frac{dx_k}{dt} + 2P_k W_k^0 x_k + y = 0 , \qquad k = 1, \ldots, n ,$$

$$F\frac{dy}{dt} = \frac{Q_0}{H + Z_0} y + \sum_1^n f_k x_k , \tag{2.4.11}$$

Our problem is to obtain conditions on F ensuring stability for this system. To describe the final result let us denote

$$a_k = \frac{2P_k W_k^0}{L_k}, \qquad b_k = \frac{H+Z_0}{Q_0}\cdot\frac{f_k}{L_k}, \qquad k=1,\ldots,n.$$

Consider a matrix with entries c_{ik} defined by

$$c_{kk} = a_k - b_k, \qquad c_{ik} = -b_k \quad (i \neq k).$$

This matrix has $n-1$ positive eigenvalues denoted by β_j and if we assume

$$a_1 < a_2 < \ldots < a_{n-1} < a_n$$

then

$$a_1 < \beta_1 < a_2 < \beta_2 < \ldots < a_{n-1} < \beta_{n-1} < a_n.$$

This assertion is proved in Appendix A.
We will prove that if

$$F > \frac{Q_0}{g(H+Z_0)}\cdot\frac{\prod_1^{n-1}\beta_k}{\prod_1^n a_k} \tag{2.4.12}$$

then the linear system (2.4.11) is exponentially stable. For $n=1$ this condition reduces to the one obtained by Thoma in 1910 (see for instance, the book of Jaeger, 1977).

To obtain the result, we consider the Liapunov function (A.Halanay, 1986):

$$V(y,x_1,\ldots,x_n) = \frac{1}{2}F\left(\frac{dy}{dt}\right)^2 + \frac{1}{2}g\left(\sum_1^n \frac{f_k}{L_k} - \alpha_0\frac{Q}{H+Z_0}\right)y^2 +$$

$$+\frac{1}{2}\sum_{k=1}^{n-1}\frac{\alpha_k}{D_k F}\left(\sum_{j=1}^n B_k^j f_j x_j\right)^2 \tag{2.4.13}$$

where α_0, α_1, ..., α_{n-1}, D_1, ..., D_{n-1}, B_k^j are defined in Appendices D, E, F.

Since

$$\alpha_0 < \frac{H+Z_0}{Q_0}\sum_{k=1}^{n}\frac{f_k}{L_k}, \qquad \frac{\alpha_k}{D_k} > 0, \quad k=1,\dots,n-1$$

(see Appendices E, F), we see that $V(y, x_1, \dots, x_n) \geq 0$. Let us find out where V can be zero: if $V = 0$ we must have $y = 0$, $\frac{dy}{dt} = 0$, $\sum_1^n f_k x_k = 0$ and, moreover $\sum_{k=1}^{n} B_j^k f_k x_k = 0$.

From the result presented in Appendix G, we deduce that $x_1 = x_2 = \dots = x_n = 0$, hence the Liapunov function is positive definite.

A straightforward but tedious computation which makes use of the facts that

$$\sum_{j=1}^{n} b_j B_i^j = D_i, \qquad a_k B_i^k - \sum_{j=1}^{n} b_j B_i^j = \beta_i B_i^k$$

leads to

$$\frac{dV}{dt} = -g\left(\frac{dy}{dt}\right)^2\left[\alpha_0 F - \frac{Q_0}{g(H+Z_0)}\right] -$$

$$-2g\frac{dy}{dt}\sum_{i=1}^{n-1}\alpha_i\sum_{k=1}^{n}B_i^k f_k x_k -$$

$$-g\sum_{i=1}^{n-1}\frac{\alpha_i\beta_i}{D_i F}\left(\sum_{k=1}^{n}B_i^k f_k x_k\right)^2 \tag{2.4.14}$$

It follows that $-\frac{1}{g}\frac{dV}{dt}$ is a quadratic form with respect to $\frac{dy}{dt}$, $\sum_{k=1}^{n} B_i^k f_k x_k$ $(i = 1, \dots, n-1)$ with the matrix

$$\begin{pmatrix} \alpha_0 F - \dfrac{Q_0}{g(H+Z_0)} & \alpha_1 & \dots & \alpha_{n-1} \\ \alpha_1 & \dfrac{\alpha_1\beta_1}{D_1 F} & \dots & 0 \\ \vdots & \vdots & & \vdots \\ \alpha_{n-1} & & & \dfrac{\alpha_{n-1}\beta_{n-1}}{D_{n-1}F} \end{pmatrix}$$

We will check the positive definiteness of this matrix using the Sylvester condition. We have

$$\Delta_1 = \alpha_0 F - \frac{Q_0}{g(H+Z_0)}, \qquad \Delta_2 = \frac{\alpha_1\beta_1}{D_1 F}\Delta_1 - \alpha_1^2,$$

$$\Delta_3 = \frac{\alpha_2\beta_2}{D_2 F}\Delta_2 - \frac{\alpha_1\beta_1}{D_1 F}\alpha_2^2, \dots,$$

$$\Delta_{l+1} = \frac{\alpha_l\beta_l}{D_l F}\Delta_l - \frac{\alpha_1\beta_1}{D_1 F}\cdots\frac{\alpha_{l-1}\beta_{l-1}}{D_{l-1}F}\alpha_l^2$$

From here we deduce

$$\frac{\Delta_{l+1}}{\frac{\alpha_1\beta_1}{D_1 F}\cdots\frac{\alpha_{l-1}\beta_{l-1}}{D_{l-1}F}\frac{\alpha_l\beta_l}{D_l F}} =$$

$$= \frac{\Delta_l}{\frac{\alpha_1\beta_1}{D_1 F}\cdots\frac{\alpha_{l-1}\beta_{l-1}}{D_{l-1}F}} - \frac{\alpha_l D_l F}{\beta_l}$$

and

$$\frac{\Delta_2}{\frac{\alpha_1\beta_1}{D_1 F}} = \Delta_1 - \frac{\alpha_1 D_1 F}{\beta_1}.$$

These relations lead to

$$\frac{\Delta_l}{\frac{\alpha_1\beta_1}{D_1 F}\cdots\frac{\alpha_{l-1}\beta_{l-1}}{D_{l-1}F}} =$$

$$= \left(\alpha_0 - \frac{\alpha_1 D_1}{\beta_1} - \dots - \frac{\alpha_{l-1}D_{l-1}}{\beta_{l-1}}\right)F - \frac{Q_0}{g(H+Z_0)}$$

and the Sylvester conditions read

$$\left(\alpha_0 - \frac{\alpha_1 D_1}{\beta_1} - \dots - \frac{\alpha_{l-1}D_{l-1}}{\beta_{l-1}}\right)F - \frac{Q_0}{g(H+Z_0)} > 0, \quad l \geq 2.$$

It is thus enough to check thet

$$\left(\alpha_0 - \sum_{k=1}^{n-1}\frac{\alpha_k D_k}{\beta_k}\right)F > \frac{Q_0}{g(H+Z_0)}$$

Taking into account the fact that

$$\alpha_0 - \sum_{k=1}^{n-1} \frac{\alpha_k D_k}{\beta_k} = \frac{\prod_1^n a_k}{\prod_1^{n-1} \beta_k} \qquad \text{(Appendix F)}$$

the following stability condition is obtained

$$F > F_{cr} = \frac{Q_0}{g(H+Z_0)} \cdot \frac{\prod_1^{n-1} \beta_k}{\prod_1^n a_k}. \tag{2.4.15}$$

To check asymptotic stability, we use again Theorem 2.5 (Corollary 2.1). The set where $\frac{dV}{dt} = 0$ is defined by

$$\frac{dy}{dt} = 0, \qquad \sum_{k=1}^{n} B_i^k f_k x_k = 0, \quad i = 1, \dots, n-1,$$

hence

$$\frac{Q_0}{H+Z_0} y + \sum_{k=1}^{n} f_k x_k = 0, \quad \sum_{k=1}^{n} B_i^k f_k x_k = 0, \quad i = 1, \dots, n-1.$$

We have already discussed the system for x_k and we have deduced that x_k are constant if y is constant. Since x_k are constant they also satisfy also

$$a_k x_k + \frac{1}{L_k} y = 0, \quad k = 1, \dots, n$$

hence

$$\sum_{k=1}^{n} f_k x_k = -\left(\sum_{k=1}^{n} \frac{f_k}{a_k L_k}\right) y.$$

Therefore

$$\left[\frac{Q_0}{H+Z_0} - \sum_{k=1}^{n} \frac{f_k}{a_k L_k}\right] y = 0.$$

A direct computation shows

$$\frac{Q_0}{H+Z_0} - \sum_{k=1}^{n} \frac{f_k}{a_k L_k} = \frac{Q_0}{H+Z_0} - \sum_{k=1}^{n} \frac{f_k W_k^0}{2P_k (W_k^0)^2} =$$

$$= \frac{Q_0}{H+Z_0} + \frac{1}{2Z_0}\sum_{k=1}^{n} f_k W_k^0 = \frac{Q_0}{H+Z_0} + \frac{Q_0}{2Z_0} =$$

$$= \frac{Q_0(H+3Z_0)}{2Z_0(H+Z_0)} \neq 0.$$

We deduce that $y = 0$ hence $x_k = 0$, $k = 1, \dots, n$ and the proof is completed.

3. Finally, we describe another application to a scheme with two tunnels feeding one surge tank and with one intermediary intake shaft on each of these tunnels. The mathematical model is

$$\frac{L_1}{g}\frac{dW_1}{dt} + P_1W_1^2 + Z_1 = 0, \qquad \frac{L_2}{g}\frac{dW_2}{dt} + P_2W_2^2 + Z_2 = 0,$$

$$F_1\frac{dZ_1}{dt} = f'(W_1 - W'), \qquad F_2\frac{dZ_2}{dt} = f''(W_2 - W''),$$

$$\frac{L'}{g}\frac{dW'}{dt} + P'(W')^2 + Z - Z_1 = 0,$$

$$\frac{L''}{g}\frac{dW''}{dt} + P''(W'')^2 + Z - Z_2 = 0,$$

$$F\frac{dZ}{dt} + Q_T = f'W' + f''W''$$

$$\eta g(H+Z)Q_T = N \tag{2.4.16}$$

Here f', f'' are the areas of the cross–sections of the tunnels, F_1, F_2 the areas of the horizontal cross–sections of the intermediate shaft, F – the area of the horizontal cross–section of the main surge tank, L_1 – the length of the tunnel from the reservoir to the intermediate shaft, W_1 – the flow velocity in this section of the tunnel, Z_1 the water level in the intermediate shaft above reservoir level; L_2, W_2, Z_2 are defined in the same way for the second tunnel and the second intermediate shaft; L' is the length of the tunnel from the intermediate shaft to the main surge tank, W' – the velocity of water flow in this section of the tunnel, L'', W''

are defined in the same way for the second tunnel; Z is the water level in the main surge tank above reservoir level; Q_T is the instantaneous flow required by the turbine. The head loss in tunnel has been written down in the natural case of positive velocity. Local head losses at the surge tank have been neglected.

The coefficients P_1, P_2, P', P'' for head losses are given by

$$P' = \frac{\lambda' L'}{2gd'}, \qquad P_1 = \frac{\lambda' L_1}{2gd'}, \qquad P'' = \frac{\lambda'' L''}{2gd''}, \qquad P_2 = \frac{\lambda'' L_2}{2gd''}$$

and we see that

$$\frac{P'}{L'} = \frac{P_1}{L_1}, \qquad \frac{P''}{L''} = \frac{P_2}{L_2}.$$

The steady-state (equilibrium) is computed from the relations

$$Z_1^0 = -P_1(W_1^0)^2, \qquad Z_2^0 = -P_2(W_2^0)^2,$$

$$W_1^0 = W_0', \qquad W_2^0 = W_0''$$

$$Z_0 - Z_1^0 = -P'(W_0')^2, \qquad Z_0 - Z_2^0 = -P''(W_0'')^2$$

$$Q_0 = f'W_0' + f''W_0'', \qquad N^2 = \eta^2 g^2 (H + Z_0)^2 Q_0^2$$

We deduce that the equilibrium water level in the surge tank is computed from the equation

$$Z_0(H + Z_0)^2 + \frac{N^2}{\eta^2 g^2 \left(\frac{f'}{\sqrt{P'+P_1}} + \frac{f''}{\sqrt{P''+P_2}}\right)^2} = 0$$

and if a natural inequality is satisfied a solution exists with $Z_0 \in (-H/3, 0)$. With an abuse of notation, the linearized system around the corresponding equilibrium will be written as follows

$$\frac{L_k}{g}\frac{dW_k}{dt} + 2P_k W_k^0 W_k + Z_k = 0, \quad k = 1, 2,$$

$$F_1 \frac{dZ_1}{dt} = f'(W_1 - W'), \qquad F_2 \frac{dZ_2}{dt} = f''(W_2 - W''),$$

$$\frac{L'}{g}\frac{dW'}{dt} + 2P'W_1^0W' + Z - Z_1 = 0,$$

$$\frac{L''}{g}\frac{dW''}{dt} + 2P''W_2^0W'' + Z - Z_2 = 0,$$

$$F\frac{dZ}{dt} = f'W' + f''W'' + \frac{Q_0 Z}{H + Z_0} \qquad (2.4.17)$$

To write down the Liapunov function, we denote by $\beta > 0$ the positive root of the equation

$$\left(x - \frac{2P_1W_1^0}{L_1}\right)\left(x - \frac{2P_2W_2^0}{L_2}\right) +$$

$$+\frac{H + Z_0}{Q_0}\left[\frac{f'}{L' + L_1}\left(x - \frac{2P_2W_2^0}{L_2}\right) + \frac{f''}{L'' + L_2}\left(x - \frac{2P_1W_1^0}{L_1}\right)\right] = 0$$

Since the second root of this equation is negative, we have

$$\beta > \frac{2P_1W_1^0}{L_1} + \frac{2P_2W_2^0}{L_2} - \frac{H + Z_0}{Q_0}\left(\frac{f'}{L' + L_1} + \frac{f''}{L'' + L_2}\right)$$

and it is also easy to see that

$$\beta < \frac{2P_1W_1^0}{L_1} + \frac{2P_2W_2^0}{L_2}$$

Denote

$$B = \frac{L'}{L' + L_1}\left(\beta - \frac{2P_1W_1^0}{L_1}\right), \qquad C = \frac{L''}{L'' + L_2}\left(\beta - \frac{2P_2W_2^0}{L_2}\right)$$

and note that $BC < 0$.

The Liapunov function is (M.Popescu and A.Halanay, 1984)

$$V(W_1, W_2, Z_1, Z_2, W', W'', Z) = \frac{1}{2}F\left(\frac{dZ}{dt}\right)^2 + \frac{1}{2}F_1\left(\frac{dZ_1}{dt}\right)^2 +$$

$$+\frac{1}{2}F_2\left(\frac{dZ_2}{dt}\right)^2 + \frac{gf'}{2L'}(Z - Z_1)^2 + \frac{gf''}{2L''}(Z - Z_2)^2 +$$

$$+\frac{gf'}{2L_1}Z_1^2+\frac{gf''}{2L_2}Z_2^2+\frac{gQ_0}{2(H+Z_0)}\left(\frac{2P_1W_1^0}{L_1}+\frac{2P_2W_2^0}{L_2}-\beta\right)Z^2+$$

$$+\frac{\mu^2}{2}\left[Bf'\left(W'+\frac{L_1}{L'}W_1\right)+Cf''\left(W''+\frac{L_2}{L''}W_2\right)\right]^2 \qquad (2.4.18)$$

where μ^2 will be chosen in the below.

Let us note that $\frac{dZ}{dt}, \frac{dZ_1}{dt}, \frac{dZ_2}{dt}$ are linear forms in the state variables hence V is a quadratic form. It is directly seen that $V \geq 0$; if $V = 0$ then $Z = Z_1 = Z_2 = 0$; next $\frac{dZ_1}{dt} = \frac{dZ_2}{dt} = 0$ imply $W_1 = W'$, $W_2 = W''$ and $\frac{dZ}{dt} = 0$ implies $f'W' + f''W'' = 0$. It follows that $W' = W'' = 0$ ($BC < 0$; we have excluded here the special case of perfect symmetry when $\frac{P_1W_1^0}{L_1} = \frac{P_2W_2^0}{L_2}$). We finally deduce that V is a positive definite quadratic form.

After a direct calculation we obtain that $-\frac{1}{g}\frac{dV}{dt}$ can be written as a quadratic form in the arguments $\frac{dZ}{dt}, \frac{dZ_1}{dt}, \frac{dZ_2}{dt}$, $Bf'\left(W'+\frac{L_1}{L'}W_1\right)+ Cf''\left(W''+\frac{L_2}{L''}W_2\right)$ with matrix (a_{ij}), $i,j = 1,\ldots,4$, where

$$a_{11}=\left(\frac{2P_1W_1^0}{L_1}+\frac{2P_2W_2^0}{L_2}-\beta\right)F-\frac{Q_0}{g(H+Z_0)},$$

$$a_{22}=\frac{2P_1W_1^0}{L_1}F_1,\quad a_{33}=\frac{2P_2W_2^0}{L_2}F_2,\quad a_{12}=-\frac{BL_1F_1}{2L'},$$

$$a_{13}=-\frac{CL_2F_2}{2L''},\quad a_{23}=0,\quad a_{14}=\frac{1}{2}(1+\mu^2DF),$$

$$a_{24}=\frac{\mu^2D}{2}\frac{L_1F_1}{L_1+L'},\quad a_{34}=\frac{\mu^2D}{2}\frac{L_2F_2}{L_2+L''},\quad a_{44}=\mu^2\beta.$$

Here

$$D=-\left(\beta-\frac{2P_1W_1^0}{L_1}\right)\left(\beta-\frac{2P_2W_2^0}{L_2}\right)=$$

$$=\frac{H+Z_0}{Q_0}\left(\frac{Bf'}{L'}+\frac{Cf''}{L''}\right)>0$$

The determinant of this matrix may be written as

$$\det\begin{pmatrix} A & b \\ b^* & \mu^2 D\frac{\beta}{D}\end{pmatrix} = (\det A)\left(\mu^2 D\frac{\beta^2}{D} - b^*A^{-1}b\right),$$

where we denoted

$$A = \begin{pmatrix} a_{11} & a_{12} & a_{13} \\ a_{12} & a_{22} & 0 \\ a_{13} & 0 & a_{33}\end{pmatrix}, \qquad b = \begin{pmatrix} a_{14} \\ a_{24} \\ a_{34}\end{pmatrix}$$

The Sylvester conditions may be written as $a_{11} > 0$, $a_{11}a_{22} - a_{12}^2 > 0$, $\det A > 0$, $\mu^2 D\frac{\beta}{D} - b^*A^{-1}b > 0$.

Denoting by A_{ij} the elements of the inverse matrix A^{-1} and $\mu^2 D = y$ the inequality

$$y\frac{\beta}{D} - b^*A^{-1}b > 0$$

can be satisfied by a proper choice of y if

$$\frac{2\beta}{D} > A_{11}F + A_{12}\frac{L_1F_1}{L_1 + L'} + A_{13}\frac{L_2F_2}{L_2 + L''},$$

$$\left(\frac{2\beta}{D} - A_{11}F - A_{12}\frac{L_1F_1}{L_1 + L'} - A_{13}\frac{L_2F_2}{L_2 + L''}\right)^2 -$$

$$-A_{11}\left[A_{11}F^2 + 2A_{12}\frac{L_1F_1F}{L_1 + L'} + 2A_{13}\frac{L_2F_2F}{L_2 + L''} + 2A_{33}\frac{L_1F_1}{L_1 + L'}\cdot\right.$$

$$\left.\cdot\frac{L_2F_2}{L_2 + L''} + A_{22}\left(\frac{L_1F_1}{L_1 + L'}\right)^2 + A_{33}\left(\frac{L_2F_2}{L_2 + L''}\right)^2\right] > 0$$

After rather tedious discussion, we obtain the conditions

$$\left(\frac{D}{2} + \frac{a_{22}a_{33}}{F_1F_2}\right)F > \frac{\beta Q_0}{g(H + Z_0)} + \beta\left(\frac{a_{12}^2}{a_{22}} + \frac{a_{13}^2}{a_{33}}\right) -$$

$$-\frac{D}{2}\left(\frac{a_{12}}{a_{22}}\frac{L_1F_1}{L_1 + L'} + \frac{a_{13}}{a_{33}}\frac{L_2F_2}{L_2 + L''}\right),$$

$$\frac{a_{22}a_{33}}{F_1F_2}F > \beta\frac{Q_0}{g(H+Z_0)} + \left(\frac{L_1B}{L'}\right)^2\frac{a_{22}}{4\beta} + \left(\frac{L_2C}{L''}\right)^2\frac{a_{33}}{4\beta},$$

$$\left(D + \frac{a_{22}a_{33}}{F_1F_2}\right)F > \beta\frac{Q_0}{g(H+Z_0)} + \beta\left(\frac{a_{12}^2}{a_{22}} + \frac{a_{13}^2}{a_{33}}\right).$$

Denote now

$$F_{cr,0} = \frac{\beta Q_0 F_1 F_2}{4g(H+Z_0)P_1W_1^0P_2W_2^0}$$

The quadratic equation for β leads to quadratic equation for $F_{cr,0}$. Denote further

$$\beta_1 = \frac{\beta L_1}{2P_1W_1^0}, \qquad \beta_2 = \frac{\beta L_2}{2P_2W_2^0}, \qquad \alpha = \sqrt{\frac{\lambda''d'(L_1+L')}{\lambda'd''(L_2+L'')}};$$

the quadratic equation for β leads to quadratic equations for β_1, β_2. We can further define

$$F_{cr,1} = \frac{1}{\beta_1+\beta_2-\beta_1\beta_2}\left[F_{cr,0} + \beta_1(\beta_2-1)^2\left(\frac{L_1}{L'+L_1}\right)^2\alpha\frac{F_1}{4} + \right.$$

$$\left. +\beta_2(\beta_1-1)^2\left(\frac{L_2}{L''+L_2}\right)^2\frac{1}{\alpha}\frac{F_2}{4}\right],$$

$$F_{cr,2} = \frac{2}{1+\beta_1+\beta_2-\beta_1\beta_2}\left[F_{cr,0} + (\beta_2-1)^2\left(\frac{L_1}{L'+L_1}\right)^2\alpha\frac{F_1}{4} + \right.$$

$$\left. +(\beta_1-1)^2\left(\frac{L_2}{L''+L_2}\right)^2\frac{1}{\alpha}\frac{F_2}{4}\right],$$

$$F_{cr,3} = F_{cr,0} + \frac{1}{\beta_2}(\beta_2-1)^2\left(\frac{L_1}{L'+L_1}\right)^2\frac{F_1}{4} +$$

$$+\frac{1}{\beta_1}(\beta_1-1)^2\left(\frac{L_2}{L''+L_2}\right)^2\frac{F_2}{4}.$$

We note that

$$\beta_1 + \beta_2 > 1 + \beta_1\beta_2 , \qquad (\beta_1 - 1)(\beta_2 - 1) < 0$$

If $\alpha < 1$ we have also

$$\alpha < \beta_1 < 1 < \beta_2 < \frac{1}{\alpha}$$

The analysis based on the Liapunov function defined by (2.4.18) leads to the stability condition

$$F > F_{cr} = \max\{F_{cr,1} , F_{cr,2} , F_{cr,3}\} \tag{2.4.19}$$

The coefficient α is a measure of asymmetry of the scheme. If $\alpha = 1$, the scheme is symmetric, $\beta_1 = \beta_2 = 1$; if α is close to 1, it follows that β_1 and β_2 are close to 1, the values $F_{cr,1}$, $F_{cr,2}$, $F_{cr,3}$ are close to $F_{cr,0}$ which is essentially the value of Thoma

$$F_{Th} = \frac{Q_0}{g(H + Z_0)} \cdot \frac{L}{2PW_0} = \frac{fd}{\lambda(H + Z_0)}$$

Let us also note that in the limit cases of strong asymmetry, when one of the tunnels has little significance, the critical area obtained is again close to Thoma's , and it is becoming closer and closer when the intermediate shaft has small section or is situated near the reservoir.

Appendix A

Consider the polynomial

$$\Pi(\lambda) = \det \begin{pmatrix} \lambda + b_1 - a_1 & b_2 & \dots & b_n \\ b_1 & \lambda + b_2 - a_2 & \dots & b_n \\ \vdots & \vdots & & \vdots \\ b_1 & b_2 & \dots & \lambda + b_n - a_n \end{pmatrix}$$

A direct computation shows that

$$\Pi(a_1) = b_1 \prod_{j=2}^{n} (a_1 - a_j)$$

and in general $\Pi(a_k) = b_k \prod_{j\neq k}(a_k - a_j)$. We assume that $a_1 < a_2 < \ldots < a_n$. Then for $j < k$, we have $a_k - a_j > 0$ while for $j > k$ we have $a_k - a_j < 0$; the number of negative factors is $n - k$ hence

$$\Pi(a_k) = b_k \prod_{j\neq k} |a_k - a_j|(-1)^{n-k}.$$

We deduce that $\Pi(a_n) > 0$, $\Pi(a_{n-1}) < 0$, ... ; it is therefore proved that there exist β_j such that $\Pi(\beta_j) = 0$ and

$$a_1 < \beta_1 < a_2 < \beta_2 < \ldots < a_{n-1} < \beta_{n-1} < a_n$$

Appendix B

By using the explicit formulae for a_k, b_k we have

$$\Pi(0) = \frac{1}{\prod_1^n L_k W_k^0} \cdot \begin{vmatrix} \frac{H+Z_0}{Q_0} f_1 W_1^0 - 2P_1(W_1^0)^2 & \ldots & \frac{H+Z_0}{Q_0} f_n W_n^0 \\ \frac{H+Z_0}{Q_0} f_1 W_1^0 & \ldots & \frac{H+Z_0}{Q_0} f_n W_n^0 \\ \vdots & & \vdots \\ \frac{H+Z_0}{Q_0} f_1 W_1^0 & \ldots & \frac{H+Z_0}{Q_0} f_n W_n^0 - 2P_n(W_n^0)^2 \end{vmatrix}$$

where $-P_k W_k^0 = Z_0$. Performing elementary calculations, we obtain

$$\Pi(0) = \frac{(H+3Z_0)(2Z_0)^{n-1}}{\prod_{k=1}^n L_k W_k^0} = (-1)^{n-1} \frac{(H+3Z_0)(2|Z_0|)^{n-1}}{\prod_{k=1}^n L_k W_k^0}$$

Appendix C

We denote by γ the n-th root of Π and write

$$\Pi(\lambda) = (\lambda - \beta_1)\ldots(\lambda - \beta_{n-1})(\lambda - \gamma)$$

hence, $\Pi(0) = (-1)^{n-1}(-\gamma)\prod_{k=1}^{n-1}\beta_k$. Comparing this expression for $\Pi(0)$ with that of Appendix B, we deduce that $\gamma < 0$ ($H + 3Z_0 > 0$ because of the choice of Z_0). On the other hand

$$\sum_{k=1}^{n-1}\beta_k + \gamma = \sum_{k=1}^{n}(a_k - b_k)$$

and since $\gamma < 0$ we deduce that

$$\sum_{k=1}^{n-1}\beta_k > \sum_{k=1}^{n}(a_k - b_k)$$

Appendix D

D. We write again $\Pi(\lambda)$ and since a_k are not roots, the equation $\Pi(\lambda) = 0$ is equivalent to the equation obtained by multiplying the first column of the determinant by $\prod_{j\neq 1}(a_j - \lambda)$, the second by $\prod_{j\neq 2}(a_j - \lambda)$ and, in general, the k–th column by $\prod_{j\neq k}(a_j - \lambda)$, after writing

$$\Pi(\lambda) = \det\begin{pmatrix} \lambda + b_1 - a_1 & b_2 & \dots & b_n \\ a_1 - \lambda & \lambda - a_2 & \dots & 0 \\ \vdots & \vdots & & \vdots \\ a_1 - \lambda & 0 & \dots & \lambda - a_n \end{pmatrix}$$

The "equivalent" equation reduces after simple manipulations to

$$\prod_{j=1}^{n}(a_j - \lambda) - \sum_{k=1}^{n} b_k \prod_{j\neq k}(a_j - \lambda) = 0.$$

Since β_k are roots of this equation, we deduce that

$$\prod_{j=1}^{n}(a_j - \beta_l) - \sum_{k=1}^{n} b_k \prod_{j\neq k}(a_j - \beta_l) = 0, \quad l = 1, \dots, n-1.$$

The eigenvector associated to the eigenvalue β_l has coordinates B_l^k satisfying

$$\begin{pmatrix} a_1 - b_1 & -b_2 & \dots & -b_n \\ -b_1 & a_2 - b_2 & \dots & -b_n \\ \vdots & \vdots & & \vdots \\ -b_1 & -b_2 & \dots & a_n - b_n \end{pmatrix}\begin{pmatrix} \beta_l^1 \\ \beta_l^2 \\ \vdots \\ \beta_l^n \end{pmatrix} = \begin{pmatrix} B_l^1 \\ B_l^2 \\ \vdots \\ B_l^n \end{pmatrix}$$

We see directly that we can take

$$B_l^k = \prod_{i \neq k} (a_i - \beta_l)$$

Appendix E

Consider now the system

$$a_k - \alpha_0 = \sum_{i=1}^{n-1} \alpha_i \prod_{j \neq k} (a_j - \beta_i), \quad k = 1, \ldots, n$$

with unknowns $\alpha_0, \alpha_1, \ldots, \alpha_{n-1}$. By simple calculations, we deduce that

$$a_k - a_1 = (a_1 - a_k) \sum_{i=1}^{n-1} \alpha_i \prod_{j \neq 1,k} (a_j - \beta_i)$$

hence

$$-1 = \sum_{i=1}^{n-1} \alpha_i \prod_{j \neq 1,k} (a_j - \beta_i), \quad k = 2, 3, \ldots, n$$

If we subtract from k–th equation the second one, we find after simple calculations that

$$0 = (a_2 - a_k) \sum_{i=1}^{n-1} \alpha_i \prod_{j \neq 1,2,k} (a_j - \beta_i)$$

that is

$$\sum_{i=1}^{n-1} \alpha_i \prod_{j \neq 1,2,k} (a_j - \beta_i) = 0.$$

Continuing in the same way, we obtain

$$-1 = \sum_{i=1}^{n-1} \alpha_i \prod_{j \neq 1,2} (a_j - \beta_i)$$

$$0 = \sum_{i=1}^{n-1} \alpha_i \prod_{j \neq 1,2,3} (a_j - \beta_i)$$

- - - - - - - - - - - - - - - - - - - -

$$0 = \sum_{i=1}^{n-1} \alpha_i(a_n - \beta_i)$$

We have subtracted successively equations with numbers 1,2,3,..., but we could have considered any other ordering to deduce that

$$0 = \sum_{i=1}^{n-1} \alpha_i(a_k - \beta_i) \qquad \text{for all } k$$

that is we must have

$$\sum_{i=1}^{n-1} \alpha_i = 0, \qquad \sum_{i=1}^{n-1} \alpha_i\beta_i = 0.$$

We deduce further that $\sum_{i=1}^{n-1} \alpha_i\beta_i^2 = 0$ and, in general

$$\sum_{i=1}^{n-1} \alpha_i\beta_i^k = 0, \quad k = 0, 1, \ldots, n-3.$$

Taking these relations into account, we deduce that

$$-1 = (-1)^{n-2} \sum_{i=1}^{n-1} \alpha_i\beta_i^{n-2}$$

that is

$$\sum_{i=1}^{n-1} \alpha_i\beta_i^{n-2} = (-1)^{n-1}$$

As a final result, we obtained that α_i satisfy a system with determinant of Vandermonde type with the elements $\beta_1, \ldots, \beta_{n-1}$ and with free term $\mathrm{col}(0, 0, \ldots, 0, (-1)^{n-1})$. We deduce that

$$\alpha_k = -\frac{1}{\prod_{i \neq k}(\beta_i - \beta_k)}, \quad k = 1, \ldots, n-1$$

We check next

$$\sum_{i=1}^{n-1} \alpha_i\beta_i^{n-1} = (-1)^{n-1} \sum_{i=1}^{n-1} \beta_i;$$

this follows from elementary properties of determinants namely

$$\left(\sum_{i=1}^{n-1}\beta_i\right)\text{Vandermonde}(\beta_1,\ldots,\beta_{n-1}) = -\begin{vmatrix} 1 & \ldots & 1 \\ \beta_1 & \ldots & \beta_{n-1} \\ \vdots & & \vdots \\ \beta_1^{n-3} & \ldots & \beta_{n-1}^{n-3} \\ \beta_1^{n-1} & \ldots & \beta_{n-1}^{n-1} \end{vmatrix},$$

$$\begin{vmatrix} 1 & 1 & \ldots & 1 & 0 \\ \beta_1 & \beta_2 & \ldots & \beta_{n-1} & 0 \\ \vdots & \vdots & & \vdots & \vdots \\ \beta_1^{n-2} & \beta_2^{n-2} & \ldots & \beta_{n-1}^{n-2} & (-1)^{n-1} \\ \beta_1^{n-1} & \beta_2^{n-1} & \ldots & \beta_{n-1}^{n-1} & (-1)^{n-1}\sum_{i=1}^{n-1}\beta_i \end{vmatrix} = 0$$

We deduce that

$$\alpha_0 = a_k - \sum_{i=1}^{n-1}\alpha_i\prod_{j\neq k}(a_j-\beta_i) =$$

$$= a_k - \sum_{i=1}^{n-1}\alpha_i\beta_i^{n-2}(-1)^{n-2}\sum_{j\neq k}a_j + (-1)^n\sum_{i=1}^{n-1}\alpha_i\beta_i^{n-1}$$

and finally

$$\alpha_0 = \sum_{j=1}^{n}a_j - \sum_{i=1}^{n-1}\beta_i\,.$$

Since $\sum_{j=1}^{n}a_j > \sum_{i=1}^{n-1}\beta_i$ we deduce that $\alpha_0 > 0$.
We have also

$$\sum_{i=1}^{n-1}\beta_i > \sum_{k=1}^{n}(a_k - b_k),$$

$$\sum_{k=1}^{n}a_k - \sum_{i=1}^{n-1}\beta_i < \sum_{k=1}^{n}b_k = \frac{H+Z_0}{Q_0}\sum_{k=1}^{n}\frac{f_k}{L_k}$$

hence

$$0 < \alpha_0 < \frac{H + Z_0}{Q_0} \sum_{k=1}^{n} \frac{f_k}{L_k}$$

Appendix F

Denote

$$D_i = \prod_{k=1}^{n} (a_k - \beta_i).$$

We have

$$a_k - \alpha_0 = \sum_{i=1}^{n-1} \frac{\alpha_i D_i}{a_k - \beta_i},$$

$$\alpha_0 - \sum_{i=1}^{n-1} \frac{\alpha_i D_i}{\beta_i} = a_1 \left[1 - \sum_{i=1}^{n-1} \frac{\alpha_i D_i}{\beta_i (a_1 - \beta_i)} \right],$$

$$-1 = \sum_{i=1}^{n-1} \frac{\alpha_i D_i}{(a_1 - \beta_i)(a_2 - \beta_i)}$$

and we write

$$\alpha_0 - \sum_{i=1}^{n-1} \frac{\alpha_i D_i}{\beta_i} = -a_1 a_2 \sum_{i=1}^{n-1} \frac{\alpha_i D_i}{\beta_i (\beta_i - a_1)(\beta_i - a_2)};$$

we have further

$$0 = \sum_{i=1}^{n-1} \frac{\alpha_i D_i}{(a_1 - \beta_i)(a_2 - \beta_i)(a_3 - \beta_i)}$$

and we deduce that

$$\alpha_0 - \sum_{i=1}^{n-1} \frac{\alpha_i D_i}{\beta_i} =$$

$$= -a_1 a_2 a_3 \sum_{i=1}^{n-1} \frac{\alpha_i D_i}{\beta_i (a_1 - \beta_i)(a_2 - \beta_i)(a_3 - \beta_i)}.$$

We can continue and deduce

$$\alpha_0 - \sum_{i=1}^{n-1} \frac{\alpha_i D_i}{\beta_i} =$$

$$= -a_1 a_2 \dots a_n \sum_{i=1}^{n-1} \frac{\alpha_i D_i}{\beta_i (a_1 - \beta_i) \dots (a_n - \beta_i)} =$$

$$= -a_1 a_2 \dots a_n \sum_{i=1}^{n-1} \frac{\alpha_i}{\beta_i} .$$

Since

$$\alpha_i = -\frac{1}{\prod_{k \neq i} (\beta_k - \beta_i)}$$

we have

$$\frac{\alpha_i}{\beta_i} = -\frac{1}{\beta_i \prod_{k \neq i} (\beta_k - \beta_i)} =$$

$$= \sum_{k \neq i} \frac{1}{\beta_k \prod_{j \neq k} (\beta_j - \beta_k)} - \frac{1}{\prod_{k=1}^{n-1} \beta_k}$$

We deduce

$$\sum_{i=1}^{n-1} \frac{\alpha_i}{\beta_i} = -\frac{1}{\prod_{k=1}^{n-1} \beta_k}$$

and

$$\alpha_0 - \sum_{i=1}^{n-1} \frac{\alpha_i D_i}{\beta_i} = \frac{\prod_{k=1}^{n} a_k}{\prod_{k=1}^{n-1} \beta_k}$$

Finally, let us check that $\alpha_i D_i > 0$. We have

$$\operatorname{sign} \prod_{k \neq 1} (\beta_k - \beta_i) = (-1)^{i-1},$$

$$\operatorname{sign} \alpha_i = (-1)^i, \qquad \operatorname{sign} D_i = (-1)^i.$$

Appendix G

G. We have

$$\begin{vmatrix} 1 & 1 & \dots & 1 \\ B_1^1 & B_1^2 & \dots & B_1^n \\ \vdots & \vdots & & \vdots \\ B_{n-1}^1 & B_{n-1}^2 & \dots & B_{n-1}^n \end{vmatrix} \neq 0$$

If the determinant were zero we could find λ_j not all zero such that

$$\sum_{j=1}^{n-1} \lambda_j (B_j^i - B_j^1) = 0, \quad i = 2, \dots, n,$$

hence such that

$$\sum_{j=1}^{n-1} \lambda_j B_j^i = \sum_{j=1}^{n-1} \lambda_j B_j^1$$

which is equivalent to

$$\sum_{j=1}^{n-1} \lambda_j \prod_{k \neq 1,i} (a_k - \beta_j) = 0.$$

On the other hand, we have

$$-1 = \sum_{j=1}^{n-1} \alpha_j \prod_{k \neq 1,i} (a_k - \beta_j)$$

and since this system has a unique solution $\alpha_1, \dots, \alpha_{n-1}$ we deduce that the corresponding system

$$\sum_{j=1}^{n-1} \lambda_j \prod_{k \neq 1,i} (a_k - \beta_j) = 0$$

has only the solution $\lambda_1 = \dots = \lambda_{n-1} = 0$ which proves our assertion.

References

BARBAŠIN, E.A., KRASOVSKII, N.N. (1952) *About the stability of motion in the large.* Dokl.A.N. S.S.S.R. **86**, *3*, 453 - 56 (in Russian).

BELLMAN, R. (1953) *Stability theory of differential equations.* Mc Graw Hill.

ČETAEV, N.G. (1946) *Stability of motion.* Gostehizdat (in Russian).

ESCANDE, L., DAT, J., PIQUEMAL, J. (1965) *Stabilité d'une chambre d'équilibre placée à la jonction de deux galéries alimentées par des lacs situés à la même côte.* C.R. Acad.Sci.Paris 261, 2579 - 2581.

HAHN, W. (1967) *Stability of motion.* Springer Verlag.

HALANAY, A. (1966) *Differential equations. Stability. Oscillations. Time lag.* Academic Press.

HALANAY, A. (1978) *Stabilitá.* Pitagora Editrice, Bologna.

HALANAY, A. (1986) *A Liapunov function for a problem in hydraulic engineering.* St.Cerc.Mat. **38**, *3*, 298 - 301.

HALANAY, A., POPESCU, M. (1979) *General operating hydraulic transient stability analysis for hydroelectric plants with one or two surge tanks via Liapunov function.* St.Cerc.Mec.Apl. **38**, *1*, 3 - 19 (in Roumanian).

JAEGER, CH. (1977) *Fluid Transients in Hydro-Electric Engineering Practice.* Blackie, Glasgow and London.

LA SALLE, J.P. (1968) *Stability Theory for Ordinary Differential Equations.* Journ. Differential Equations, **4**, *1*, 57 - 65.

MALKIN, I.G. (1952) *Stability of Motion.* Gostehizdat, Moscow (in Russian).

POPESCU, M., HALANAY, A. (1984) *Hydraulic stability of Hydropower stations working in complex schemes,* The Second Symposium on Fluid Motion. Stability in Hydraulic Systems with Automatic Control. Romania, Bucharest. IAHR paper A–15.

ROUCHE, N., HABETS, P., LALOY, M. (1977) *Stability Theory by Liapunov Direct Method.* Springer Verlag.

YOSHIZAWA, T. (1966) *Stability Theory by Liapunov Second Method.* The Math.Soc.of Japan, Tokyo.

CHAPTER 3

Stability Problems in Power Engineering

3.1 Stability of Synchronous Generators. Mathematical Models of Synchronous Machine

The equations describing the electromechanical transient behaviour of a synchronous machine with flux linkage variations, machine damping and transient saliency included, in the case of balanced generator operation, are the following (B.Adkins, 1962):

$$
\begin{aligned}
u_d &= -R_s i_d - \dot{\psi}_d - \omega(1+s)\psi_q \\
u_q &= -R_s i_q - \dot{\psi}_q + \omega(1+s)\psi_d \\
u_f &= R_f i_f + \dot{\psi}_f \qquad\qquad (3.1.1) \\
0 &= R_D i_D + \dot{\psi}_D \\
0 &= R_Q i_Q + \dot{\psi}_Q \\
\dot{\delta} &= \omega s\,; \qquad T\dot{s} + Ds = P_{mec} - \omega(\psi_d i_q - \psi_q i_d)
\end{aligned}
$$

together with the linear algebraic relations between currents and flux linkages (The linearity of these relations follows from the assumption that the machine is unsaturated):

$$
\begin{aligned}
\psi_d &= L_d i_d + M_{ad} i_f + M_{ad} i_D \\
\psi_f &= M_{ad} i_d + L_f i_f + M_{ad} i_D \\
\psi_D &= M_{ad} i_d + M_{ad} i_f + L_D i_D \\
\psi_q &= L_q i_q + M_{aq} i_Q \\
\psi_Q &= M_{aq} i_q + L_Q i_Q
\end{aligned}
\tag{3.1.2}
$$

If the inductances of the above equations are replaced by the corresponding synchronous, field mutual, and damper reactances

$$
x_d = \omega L_d\,, \qquad x_f = \omega L_f\,, \qquad x_D = \omega L_D\,, \qquad x_q = \omega L_q\,,
$$

$$
x_Q = \omega L_Q\,, \qquad x_{ad} = \omega M_{ad}\,, \qquad x_{aq} = \omega M_{aq}
$$

and if the transient, sub–transient and leakage reactances x'_d, x''_d, x''_q, x_σ are defined from the relations

$$
\begin{aligned}
x_\sigma &= x_d - x_{ad} = x_q - x_{aq}\,, \\
x_f &= (x_d - x_\sigma)^2/(x_d - x'_d) \\
x_D &= x_d - x'_d + (x'_d - x_\sigma)^2/(x'_d - x''_d)\,, \\
x_Q &= (x_q - x_\sigma)^2/(x_q - x''_q)
\end{aligned}
$$

then the flux linkages are expressed as follows

$$
\omega\psi_d = x_d i_d + (x_d - x_\sigma) i_f + (x_d - x_\sigma) i_D
$$

$$
\omega\psi_f = (x_d - x_\sigma) i_d + \frac{(x_d - x_\sigma)^2}{x_d - x'_d} i_f + (x_d - x_\sigma) i_D
$$

$$\omega\psi_q = x_q i_q + (x_q - x_\sigma) i_Q$$

$$\omega\psi_D = (x_d - x_\sigma) i_d + (x_d - x_\sigma) i_f + \left(x_d - x'_d + \frac{(x'_d - x_\sigma)^2}{x'_d - x''_d} \right) i_D$$

$$\omega\psi_Q = (x_q - x_\sigma) i_q + \frac{(x_q - x_\sigma)^2}{x_q - x''_q} i_Q .$$

Introducing the new variables

$$e''_d = -\frac{x_q - x''_q}{x_q - x_\sigma} \omega\psi_Q , \qquad e'_q = \frac{x_d - x'_d}{x_d - x_\sigma} \omega\psi_f$$

$$v = \omega\psi_q \sin\delta + \omega\psi_d \cos\delta \tag{3.1.3}$$

$$w = \omega\psi_q \cos\delta - \omega\psi_d \sin\delta$$

the following synchronous generator model with seven differential equations is obtained:

$$\dot{\delta} = \omega s$$

$$T\dot{s} + Ds = P_{mec} - \left[\left(\frac{1}{x''_q} - \frac{1}{x''_d} \right) (v\cos\delta - w\sin\delta)(v\sin\delta + \right.$$

$$\left. + w\cos\delta) + \frac{e''_d}{x''_q}(v\cos\delta - w\sin\delta) + \frac{e''_q}{x''_d}(v\sin\delta + w\cos\delta) \right]$$

$$T'_{d_0}\dot{e}'_q = e_f - \frac{x_d - x_\sigma}{x'_d - x_\sigma} e'_q + \frac{x_\sigma}{x''_d} \cdot \frac{x_d - x'_d}{x'_d - x_\sigma} e''_q +$$

$$+ \frac{(x''_d - x_\sigma)(x_d - x'_d)}{x''_d (x'_d - x_\sigma)} (v\cos\delta - w\sin\delta)$$

$$T''_{d0}\dot{e}''_q = e'_q - \frac{x'_d}{x''_d} e''_q + \frac{x'_d - x''_d}{x''_d} (v\cos\delta - w\sin\delta) +$$

$$+ T''_{d0} \frac{x''_d - x_\sigma}{x'_d - x_\sigma} \dot{e}'_q$$

$$T''_{q0}\dot{e}''_d = -\frac{x_q}{x''_q}e''_d - \frac{x_q - x''_q}{x''_q}(v\sin\delta + w\cos\delta)$$

$$\frac{1}{\omega}\dot{v} = -R_s\left(\frac{\sin^2\delta}{x''_q} + \frac{\cos^2\delta}{x''_d}\right)v -$$

$$-\left[1 + R_s\left(\frac{1}{x''_q} - \frac{1}{x''_d}\right)\sin\delta\cos\delta\right]w - \tag{3.1.4}$$

$$-R_s\left(\frac{e''_d\sin\delta}{x''_q} - \frac{e''_q\cos\delta}{x''_d}\right) - (u_d\cos\delta + u_q\sin\delta)$$

$$\frac{1}{\omega}\dot{w} = \left[1 - R_s\left(\frac{1}{x''_q} - \frac{1}{x''_d}\right)\sin\delta\cos\delta\right]v - R_s\left(\frac{\cos^2\delta}{x''_q} +\right.$$

$$\left. + \frac{\sin^2\delta}{x''_d}\right)w - R_s\left(\frac{e''_d\cos\delta}{x''_q} - \frac{e''_q\sin\delta}{x''_d}\right) - (u_q\cos\delta - u_d\sin\delta).$$

Here the following conventions are being employed:

$$e_f = \frac{x_d - x_\sigma}{R_f}u_f, \qquad T'_{d0} = \frac{(x_d - x_\sigma)^2}{\omega R_f(x_d - x'_d)} = \frac{L_f}{R_f}$$

$$T''_{d0} = \frac{(x'_d - x_\sigma)^2}{\omega R_D(x'_d - x''_d)}, \qquad T''_{q0} = \frac{(x_q - x_\sigma)^2}{\omega R_Q(x_q - x''_q)}$$

The fact that ω is large suggests that v and w are fast variables and that it would be possible to replace the corresponding differential equations by algebraic ones by applying the singular perturbations theory (A.Halanay and V.Drăgan, 1983). Therefore, a preliminary check of the roots for the characteristic equation of the Jacobi matrix for v and w is necessary. This characteristic equation is

$$\lambda^2 + R_s\left(\frac{1}{x''_q} + \frac{1}{x''_d}\right)\lambda + 1 + \frac{R_s^2}{x''_d x''_q} = 0$$

and the Jacobi matrix is a Hurwitz matrix. Expressing v and w from their differential equations (where $1/\omega = 0$) and substituting the result into (3.1.4), a model with five differential equations is obtained. This model has quite lengthy expressions and will not be reproduced here; it is called the model with neglected stator dynamics. Considering this model as basic for what follows and neglecting R_s the following simplified model is obtained

$$\begin{aligned}
&\dot{\delta} = \omega s \\
&T\dot{s} + Ds = P_{mec} - \left[\frac{e''_d u_q}{x''_q} - \frac{e''_q u_d}{x''_d} + \left(\frac{1}{x''_d} - \frac{1}{x''_q}\right) u_d u_q\right] \\
&T'_{do}\dot{e}'_q = e_f - \frac{x_d - x_\sigma}{x'_d - x_\sigma} e'_q + \frac{x_\sigma (x_d - x'_d)}{x''_d (x'_d - x_\sigma)} e''_q + \\
&+ \frac{(x''_d - x_\sigma)(x_d - x'_d)}{x''_d (x'_d - x_\sigma)} u_q \qquad (3.1.5) \\
&T''_{do}\dot{e}''_q = e'_q - \frac{x'_d}{x''_d} e''_q + \frac{x'_d - x''_d}{x''_d} u_q + T''_{do} \frac{x''_d - x_\sigma}{x'_d - x_\sigma} \dot{e}'_q \\
&T''_{qo}\dot{e}''_d = -\frac{x_q}{x''_q} e''_d + \frac{x_q - x''_q}{x''_q} u_d
\end{aligned}$$

This five–equations model is used very often for stability studies. It is worth mentioning that if a new variable is introduced, namely

$$\hat{e}''_q = e''_q - \frac{x''_d - x_\sigma}{x'_d - x_\sigma} e'_q$$

the model obtained from (3.1.5) will coincide with the model of Fagiuoli and Szegö (1970) for the unloaded ($P_{mec} = 0$) synchronous generator.

If e''_q and e''_d are regarded as fast variables and the time constants T''_{do} and T''_{qo} (which are at most one tenth of T'_{qo} and T) are neglected, a model with three differential equations is obtained:

$$\begin{aligned}
&\dot{\delta} = \omega s \\
&T\dot{s} + Ds = P_{mec} - \left[-\frac{e'_q u_d}{x'_d} + \left(\frac{1}{x'_d} - \frac{1}{x_q}\right) u_d u_q\right] \qquad (3.1.6)
\end{aligned}$$

$$T'_{d0}\dot{e}'_q = e_f - \frac{x_d}{x'_d}e'_q + \frac{x_d - x'_d}{x'_d}u_q$$

In this model only mechanical and field circuit transients are considered.

Quite often the time constant T'_{d0} can be neglected, especially if only mechanical transients are considered. In this case the following model is obtained:

$$\dot{\delta} = \omega s$$

$$T\dot{s} + Ds = P_{mec} - \left[-\frac{e_f u_d}{x_d} + \left(\frac{1}{x_d} - \frac{1}{x_q}\right)u_d u_q\right] \qquad (3.1.7)$$

In fact, the qualitative problems are the same for all models. As a working assumption we will consider a generator connected to an infinite bus. This assumption is obviously an idealization, but it allows for the detailed study of the behaviour of a single generator, the analysis of transient processes, and the automatic controller design. The connection to the infinite bus is modelled by the following expressions for the voltages u_d, u_q:

$$u_d = -U\sin\delta\,, \qquad u_q = U\cos\delta \qquad (3.1.8)$$

which have to be substituted in (3.1.4) – (3.1.7).

Assuming that the mechanical power P_{mec} supplied by the prime mover (the turbine) and the field voltage e_f are constant, the number of the differential equations equals the number of the unknowns. The synchronous generator has to work stable at an operation point. This means nothing else but the requirement that the system of differential equations should have a stable singular (stationary, steady–state) point.

The stability property has to be considered with respect to various disturbances, both electrical (short–circuits, consumer switching on and off) and mechanical (nonconstant active power, speed controller signals).

These disturbances will be considered as short–time, hence as generating initial conditions for perturbed motions around the singular point. In addition to this, there exists another case leading to a stability problem. The electric energy system is often subject to "manoeuvres": the mechanical power of the prime mover is modified, "loading" or "unloading" the generator. The mechanical power modification leads to a new steady state of the generator while the old steady state becomes the initial condition of a perturbed motion in the neighbourhood of the new steady state.

It can be seen from (3.1.7) that the equations of the synchronous machine have several singular points defined by

$$P_{mec} = \frac{e_f U}{x_d} \sin\delta + \left(\frac{1}{x_q} - \frac{1}{x_d}\right) U^2 \sin\delta\cos\delta$$

which leads to the following algebraic equation

$$\left(P_{mec} + \frac{U^2}{x_q} - \frac{U^2}{x_d}\right)\xi^4 - \frac{2e_f U}{x_d}\xi^3 + 2P_{mec}\xi^2 -$$

$$-\frac{2e_f U}{x_d}\xi + \left(P_{mec} - \frac{U^2}{x_q} + \frac{U^2}{x_d}\right) = 0$$

(The state variables e''_q, e''_d, e'_q are uniquely determined if δ is known; this can be easily verified when the singular perturbation model reduction is applied).

The existence of several singular points implies that the stability of a given singular point is only local. Therefore, the initial conditions of a perturbed motion should belong to the attraction domain of the singular point whose stability is studied.

Consequently, a "manoeuvre" should be performed in order to have the initial steady state located in the attraction domain of the new steady state required by the "manoeuvre".

The practical experience of the electric energy specialists lead to the representation of two kinds of transitions from a steady state regime to another (A.A.Gorev, 1950).

By *simple transition*, one understands the transient process from one steady state to another under persistent disturbances. The disturbance is not too large and the new steady state is quite close to the initial one; from the mathematical point of view this justifies the linearization around the final steady state. It is thus possible to make use of the *steady–state stability* concept.

A *complex transition* has at least two phases: the fault and the post–fault recovery (e.g. a short-circuit and the post–fault pendulation). In this case not only the disturbances are large but even the new steady-state can be very far from the initial one. The linearization is no longer possible and the *transient stability* concept has to be used.

In fact even the concept of transition shows that in both cases the same kind of stability is required – the Liapunov stability i.e., the stability with respect to motions perturbed by initial conditions. The following study will be based on the concept of Liapunov stability which is unifying the two stability concepts mentioned above.

Liapunov Functions Associated with Synchronous Machines

Let us consider the simplest model (3.1.7) with two differential equations, with u_d, u_q defined by (3.1.8) and let us associate with this model the following functions:

$$\varphi(\delta) = \frac{e_f U}{x_d}\sin\delta + U^2\left(\frac{1}{x_q} - \frac{1}{x_d}\right)\sin\delta\cos\delta - P_{mec}$$

$$V_2(\delta, s) = \frac{1}{2}\omega s^2 + \frac{1}{T}\int_{\hat{\delta}}^{\delta}\varphi(\theta)d\theta \qquad (3.1.9)$$

where $\hat{\delta}$ corresponds to a singular (steady–state) point of the system of differential equations.

It can be seen quite easily that the system (3.1.7) – (3.1.8) takes the form

$$\dot{\delta} = \frac{\partial V_2}{\partial s}, \qquad \dot{s} = -\frac{\partial V_2}{\partial \delta} - \frac{D}{T}s \qquad (3.1.10)$$

hence, it is a Hamiltonian system with dissipation term.

Consider now the model (3.1.6) with u_d, u_q defined by (3.1.8) and associate the function

$$V_3(\delta, s, e'_q) = \frac{1}{2}\omega s^2 + \frac{1}{T}\left\{\frac{1}{2}\frac{x_d}{x'_d(x_d - x'_d)}\cdot\right.$$

$$\left.\cdot\left[e'_q - \frac{x'_d}{x_d}e_f - \frac{x_d - x'_d}{x_d}U\cos\delta\right]^2 + \int_{\hat{\delta}}^{\delta}\varphi(\theta)d\theta\right\} \tag{3.1.11}$$

It can be found by straightforward computation that the system (3.1.6), (3.1.8) can be written as

$$\dot{\delta} = \frac{\partial V_3}{\partial s}, \qquad \dot{s} = -\frac{\partial V_3}{\partial \delta} - \frac{D}{T}s$$

$$\dot{e}'_q = -\frac{T}{T'_{d0}}(x_d - x'_d)\frac{\partial V_3}{\partial e'_q} \tag{3.1.12}$$

Equations (3.1.12) represent a Hamiltonian system coupled with a gradient one; the coupling takes place via the generating function V_3.

Let us note that the Hamiltonian sub–system describes the mechanical processes while the gradient sub–system describes the electrical processes. Under these circumstances an attempt is made to find a function to get the same structure for the system described by (3.1.5) and (3.1.8).

Some simple but tedious manipulations lead to the function

$$V_5(\delta, s, e'_q, e''_q, e''_d) = \frac{1}{2}\omega s^2 + \frac{1}{T}\left\{\frac{1}{2}\frac{x_d - x''_d}{(x_d - x'_d)(x'_d - x''_d)}\cdot\right.$$

$$\cdot\left[e'_q - \frac{x_d - x'_d}{x_d - x''_d}e''_q - \frac{x'_d - x''_d}{x_d - x''_d}e_f\right]^2 + \frac{1}{2}\frac{x_d}{x''_d(x_d - x''_d)}\cdot$$

$$\cdot\left[e''_d - \frac{x''_d}{x_d}e_f - \frac{x_d - x''_d}{x_d}U\cos\delta\right]^2 + \tag{3.1.13}$$

$$\left.+\frac{1}{2}\frac{x_q}{x''_q(x_q - x''_q)}\left[e''_d + \frac{x_q - x''_q}{x_q}U\cos\delta\right]^2 + \int_{\hat{\delta}}^{\delta}\varphi(\theta)d\theta\right\}$$

with the help of which the system (3.1.5), (3.1.8) takes the form

$$\dot{\delta} = \frac{\partial V_5}{\partial s}, \qquad \dot{s} = -\frac{\partial V_5}{\partial \delta} - \frac{D}{T}s,$$

$$\dot{e}'_q = -\frac{T}{T'_{do}}(x_d - x'_d)\frac{\partial V_5}{\partial e'_q} - \frac{x''_d - x_\sigma}{x'_d - x_\sigma}\frac{T}{T'_{do}}\frac{\partial V_5}{\partial e''_q}$$

$$\dot{e}''_q = -\frac{x''_d - x_\sigma}{x'_d - x_\sigma}\frac{T}{T'_{do}}\frac{\partial V_5}{\partial e'_q} - \tag{3.1.14}$$

$$-\left[\frac{T}{T''_{do}}(x'_d - x''_d) + \left(\frac{x''_d - x_\sigma}{x'_d - x_\sigma}\right)^2 \frac{T}{T'_{do}}(x_d - x'_d)\right]\frac{\partial V_5}{\partial e''_q}$$

$$\dot{e}''_d = -\frac{T}{T''_{q0}}(x_q - x''_q)\frac{\partial V_5}{\partial e''_d}$$

This system is a Hamiltonian system coupled with a gradient one, as before. Taking into account both the structure of all systems considered above and of the functions V_i generating them, the following general structure of Hamiltonian system coupled with a gradient one can be considered:

$$\dot{\delta} = \frac{\partial V}{\partial s}, \qquad \dot{s} = -\frac{\partial V}{\partial \delta} - \alpha s$$

$$\dot{x} = -B\left(\frac{\partial V}{\partial x}\right)^* \tag{3.1.15}$$

where $\frac{\partial V}{\partial x}$ is a row vector and B is a symmetric positive definite matrix. In some cases matrix B is also diagonal. For instance, if the magnetic circuit of the electrical machine has some symmetry, more precisely if $L_D = M_{ad}$ i.e., $x''_d = x_\sigma$ then matrix B in (3.1.14) becomes diagonal (E.Arie et al.,1974). This can be seen immediately by examining equations (3.1.14) and taking $x''_d = x_\sigma$. Also, if the state variable $\hat{e}''_q$, defined above and used by Fagiuoli and Szegö (1970), is introduced, the following system is obtained, with diagonal B.

$$\dot{\delta} = \frac{\partial \hat{V}_5}{\partial s}, \qquad \dot{s} = -\frac{\partial \hat{V}_5}{\partial \delta} - \frac{D}{T}s$$

$$\dot{e}'_q = -\frac{T}{T'_{d0}}(x_d - x'_d)\frac{\partial \hat{V}_5}{\partial e'_q}$$

$$\dot{\hat{e}}''_q = -\frac{T}{T''_{d0}}(x'_d - x''_d)\frac{\partial \hat{V}_5}{\partial \hat{e}''_q} \qquad (3.1.14')$$

$$\dot{e}''_d = -\frac{T}{T''_{d0}}(x_q - x''_q)\frac{\partial \hat{V}_5}{\partial e''_d}$$

where

$$\hat{V}_5(\delta, s, e'_q, \hat{e}''_q, e''_d) = V_5(\delta, s, e'_q, \hat{e}''_q + \frac{x''_d - x_\sigma}{x'_d - x_\sigma}e'_q, e''_d)$$

Thus, we found that a large class of synchronous machine models can be studied in the framework of the model (3.1.15). Concerning the function $V(\delta, s, x)$, the form of functions V_i $(i = 2,3,5)$ suggests the following form

$$V(\delta, s, x) = \frac{1}{2}\omega s^2 + \frac{1}{2}x^*Ax + \frac{1}{2}x^*c(\delta) + \frac{1}{2}c^*(\delta)x + \frac{1}{2}\chi(\delta) \qquad (3.1.16)$$

where A is symmetric positive definite. The fact that $A > 0$ shows that V can be used as a Liapunov function for stability and other qualitative studies.

Inherent Stability of Synchronous Machines

The models considered for synchronous machines do not include governors nor the dynamics of prime movers. This is why the stability property for such models represents a natural, inherent property of the machine. We state the main result in a form of the following theorem.

Theorem 3.1 *Consider the system* (3.1.15) *with* $V(s, x, \delta)$ *defined by* (3.1.16) *and assume that* $A > 0$, $B > 0$. *Let* $(0, \hat{x}, \hat{\delta})$ *be a constant solution of* (3.1.15) *(steady state, equilibrium). Define the function* $\varphi : R \longrightarrow R$ *by*

$$\varphi(\delta) = \chi'(\delta) - c^*(\delta)A^{-1}c(\delta) - [c'(\delta)]^*A^{-1}c(\delta) \qquad (3.1.17)$$

Then, if $\varphi'(\hat{\delta}) > 0$, *the equilibrium is asymptotically stable.*

Proof Let us note first that an equilibrium is defined by

$$s = 0 \ (\frac{\partial V}{\partial s} = 0), \qquad \frac{\partial V}{\partial \delta} = 0, \qquad \frac{\partial V}{\partial x} = 0$$

hence it is a critical point for V.

We compute derivative of V along solutins of the system:

$$\frac{dV}{dt}(s,x,\delta) = \frac{\partial V}{\partial \delta}(s,x,\delta)\frac{\partial V}{\partial s}(s,x,\delta) +$$

$$+\frac{\partial V}{\partial s}(s,x,\delta)\left[-\alpha s - \frac{\partial V}{\partial \delta}(s,x,\delta)\right] +$$

$$+\frac{\partial V}{\partial x}(s,x,\delta)\left[-B\left(\frac{\partial V}{\partial x}(s,x,\delta)\right)^{*}\right] =$$

$$= -\alpha\omega s^2 - \frac{\partial V}{\partial x}(s,x,\delta)B\left[\frac{\partial V}{\partial x}(s,x,\delta)\right]^{*} \leq 0.$$

Let us assume now that the critical point $(0,\hat{x},\hat{\delta})$ is a minimum for V; then for (s,x,δ) in a neighbourhood of $(0,\hat{x},\hat{\delta})$, we have

$$V(s,x,\delta) - V(0,\hat{x},\hat{\delta}) \geq 0$$

with equality only at the point $(0,\hat{x},\hat{\delta})$ if this is an isolated minimum. It is thus clear that $\mathcal{V}$ defined by

$$\mathcal{V}(s,x,\delta) = V(s,x,\delta) - V(0,\hat{x},\hat{\delta})$$

satisfies the usual conditions of the Liapunov stability theorem (see Theorem 2.3) and the equilibrium poit $(0,\hat{x},\hat{\delta})$ is stable. If the critical point is not a minimum, then the Liapunov instability theorem shows that the equilibrium is unstable.

Therefore, we must check if under the assumption made in the statement above, the point $(0,\hat{x},\hat{\delta})$ is an isolated minimum. We consider the corresponding Hessian matrix

$$\begin{pmatrix} \omega & 0 & 0 \\ 0 & A & c'(\hat{\delta}) \\ 0 & [c'(\hat{\delta})]^{*} & \frac{1}{2}\left[\chi''(\delta) + \hat{x}^{*}c''(\hat{\delta}) + [c''(\hat{\delta})]^{*}\hat{x}\right] \end{pmatrix}$$

where $(\hat{x}, \hat{\delta})$ satisfy the equations (for the critical points)

$$A\hat{x} + c(\hat{\delta}) = 0, \qquad \chi'(\hat{\delta}) + \hat{x}^* c'(\hat{\delta}) + [c'(\hat{\delta})]^* \hat{x} = 0.$$

Since $\omega > 0$, $A > 0$, the condition for an isolated minimum is

$$\det \begin{pmatrix} A & c'(\hat{\delta}) \\ [c'(\hat{\delta})]^* & \frac{1}{2}\left[\chi''(\hat{\delta}) + \hat{x}^* c''(\hat{\delta}) + [c''(\hat{\delta})]^* \hat{x}\right] \end{pmatrix} > 0$$

But $\hat{x} = -A^{-1}c(\hat{\delta})$, hence the above condition reads

$$\det \begin{pmatrix} A & c'(\hat{\delta}) \\ [c'(\hat{\delta})]^* & \frac{1}{2}\left[\chi''(\hat{\delta}) - c^*(\hat{\delta})A^{-1}c''(\hat{\delta}) - [c''(\hat{\delta})]^* A^{-1} c(\hat{\delta})\right] \end{pmatrix} > 0$$

and using the fact that $A > 0$, this condition turns out to be equivalent to

$$\frac{1}{2}\left[\chi''(\hat{\delta}) - c^*(\hat{\delta})A^{-1}c''(\hat{\delta}) - [c''(\hat{\delta})]^* A^{-1} c(\hat{\delta}) - \right.$$

$$\left. - [c'(\hat{\delta})]^* A^{-1} c'(\hat{\delta})\right] > 0$$

that is

$$\chi''(\hat{\delta}) - c^*(\hat{\delta})A^{-1}c''(\hat{\delta}) - [c''(\hat{\delta})]^* A^{-1} c(\hat{\delta}) -$$

$$-2[c'(\hat{\delta})]^* A^{-1} c'(\hat{\delta}) > 0$$

Taking definition of φ into account, this is exactly $\varphi'(\hat{\delta}) > 0$.

The stability is asymptotic. The derivative $\frac{dV}{dt}$ equals zero only on the set defined by

$$s = 0, \qquad \frac{\partial V}{\partial x}(s, x, \delta) = 0;$$

from the second equality it follows that on this set x is constant and since $s = 0$ it follows that δ is constant as well. According to the theorem of Barbašin – Krasovskii – La Salle (Theorem 2.5) all solutions starting in a bounded set $V(s,x,\delta) < c$ will tend to the largest invariant set contained in the set defined by the above equalities, that is to the set of constant solutions. The equilibrium $(0,\hat{x},\hat{\delta})$ is isolated and Liapunov stable, hence, solutions starting in a sufficiently small neighbourhood of this solution cannot tend to a different equilibrium.

Asymptotic stability is thus proved.

Stability of the Synchronous Machine with Prime Mover and Speed Governor

We consider linearized equations for the prime mover and speed governor

$$\dot{y} = Hy + bs\,, \qquad \zeta = q^* y - \beta s$$

where the output ζ acts as a correction of the mechanical power of the prime mover

$$P_{mec} = P_{mec_0} + \zeta$$

Coupling of the prime mover and the governor to the machine leads to

$$\begin{aligned}
\dot{\delta} &= \frac{\partial V}{\partial s}(s,x,\delta)\,,\\
\dot{s} &= -\frac{\partial V}{\partial \delta}(s,x,\delta) - (\alpha+\beta)s + q^* y\,,\\
\dot{x} &= -B\left(\frac{\partial V}{\partial x}(s,x,\delta)\right)^*\,,\\
\dot{y} &= Hy + bs
\end{aligned} \tag{3.1.18}$$

Theorem 3.2 *Assume as above* $A > 0$, $B > 0$; *assume also that* H *is a Hurwitz matrix,* (H,b) *a controllable pair and* (q^*,H) *an observable pair. Assume, as in Theorem* 3.1 *that* $\varphi'(\hat{\delta}) > 0$. *Then, if the frequency domain condition*

$$\alpha + \beta - \Re q^*(\imath\lambda I - H)^{-1} b > 0 \tag{3.1.19}$$

is satisfied for all $\lambda \in R$, *the equilibrium* $s = 0$, $x = \hat{x}$, $\delta = \hat{\delta}$, $y = 0$ *is asymptotically stable.*

Proof We will use the Yakubovich – Kalman – Popov lemma; the frequency domain condition (3.1.19) leads to existence of γ, w, G such that

$$H^*G + GH = -ww^*,$$

$$\omega(\alpha + \beta) = \gamma^2,$$

$$-\frac{\omega}{2}q - Gb = \gamma w$$

Since H is a Hurwitz matrix it is clear that $G \geq 0$.
Consider the function $\tilde{V}$ defined by

$$\tilde{V}(s, x, \delta, y) = V(s, x, \delta) + y^*Gy$$

The derivative solutions of the system is

$$\frac{d\tilde{V}}{dt}(s, x, \delta, y) = \frac{\partial V}{\partial s}(s, x, \delta)\left[-\frac{\partial V}{\partial \delta}(s, x, \delta) - (\alpha + \beta)s + q^*y\right] +$$

$$+\frac{\partial V}{\partial \delta}(s, x, \delta)\frac{\partial V}{\partial s}(s, x, \delta) + \frac{\partial V}{\partial x}(s, x, \delta)\left[-B\left(\frac{\partial V}{\partial x}(s, x, \delta)\right)^*\right] +$$

$$+y^*G(Hy + bs) + (y^*H^* + sb^*)Gy =$$

$$= -\frac{\partial V}{\partial x}(s, x, \delta)B\left(\frac{\partial V}{\partial x}(s, x, \delta)\right)^* - (\alpha + \beta)\omega s^2 + \omega s q^* y +$$

$$+y^*(GH + H^*G)y + y^*Gbs + sb^*Gy =$$

$$= -\frac{\partial V}{\partial x}(s, x, \delta)B\left(\frac{\partial V}{\partial x}(s, x, \delta)\right)^* - \gamma^2 s^2 + \omega s q^* y - y^*ww^*y -$$

$$-\frac{1}{2}y^*q\omega s - \frac{1}{2}\omega s q^* y - \gamma s y^* w - \gamma s w^* y =$$

$$= -\frac{\partial V}{\partial x}(s,x,\delta)B\left[\frac{\partial V}{\partial x}(s,x,\delta)\right]^* - (\gamma s + w^* y)^2 \leq 0$$

From the last inequality it follows that

$$\tilde{V}(s(t),x(t),\delta(t),y(t)) - \tilde{V}(0,\hat{x},\hat{\delta},0) \leq$$

$$\leq \tilde{V}(s(0),x(0),\delta(0),y(0)) - \tilde{V}(0,\hat{x},\hat{\delta},0).$$

On the other hand, since $G \geq 0$, we have

$$\tilde{V}(s(t),x(t),\delta(t),y(t)) - \tilde{V}(0,\hat{x},\hat{\delta},0) \geq$$

$$\geq V(s(t),x(t),\delta(t)) - V(0,\hat{x},\hat{\delta}).$$

Under assumptions of the theorem, $(0,\hat{x},\hat{\delta})$ is an isolated minimum and

$$V(s(t),x(t),\delta(t)) - V(0,\hat{x},\hat{\delta}) \geq$$

$$\geq K\left(\sqrt{(s(t))^2 + |x(t)-\hat{x}|^2 + (\delta(t)-\hat{\delta})^2}\right)$$

where $K(\rho)$ is a Massera function. We deduce that

$$\sqrt{(s(t))^2 + |x(t)-\hat{x}|^2 + (\delta(t)-\hat{\delta})^2} \leq$$

$$\leq \hat{K}\left(\sqrt{(s(0))^2 + |x(0)-\hat{x}|^2 + (\delta(0)-\hat{\delta})^2 + |y(0)|^2}\right)$$

where $\hat{K}(\rho)$ is again a Massera function.
Since, in particular, this inequality holds for $s(t)$ and we have

$$\dot{y}(t) = Hy(t) + bs(t)$$

with H a Hurwitz matrix, we deduce that we have the same type of inequality also for y and it is clear that the equilibrium $(0,\hat{x},\hat{\delta},0)$ is stable. To prove that it is asymptotically stable we will use the Barbašin

– Krasovskii – La Salle theorem (Theorem 2.5); we have $\frac{d\tilde{V}}{dt} = 0$ on the set

$$\frac{\partial V}{\partial x}(s, x, \delta) = 0\,, \qquad \gamma s + w^* y = 0$$

The first equality shows that on this set x must be constant and since this equality reads

$$Ax + c(\delta) = 0\,, \qquad A > 0\,,$$

we deduce that δ must be constant as well, hence s must be zero and therefore $w^* y = 0$. Since $\dot{s} = 0$, we have

$$q^* y = \frac{\partial V}{\partial \delta}(s, x, \delta).$$

Therefore, $q^* y$ must be constant. On the other hand, since $s = 0$, we have $y(t) = e^{Ht} y_0$, $q^* y(t) = q^* e^{Ht} y_0$ must be constant. It follows that $q^* H e^{Ht} y_0 \equiv 0$, $q^* H^2 e^{Ht} y_0 \equiv 0$, ..., $q^* H^n e^{Ht} y_0 \equiv 0$ and for $t = 0$ we have

$$q^* H y_0 = 0\,, \;\; q^* H^2 y_0 = 0\,, \;\; \ldots, \;\; q^* H^n y_0 = 0.$$

If the pair (q^*, H) is observable, we deduce that $H y_0 = 0$ and since H is invertible, $y_0 = 0$, hence $y(t) \equiv 0$.

We have proved that the only invariant set contained in the set where $\frac{d\tilde{V}}{dt} = 0$ consists of equilibria with $s = 0$, $y = 0$ and since $(0, \hat{x}, \hat{\delta}, 0)$ is stable it must be asymptotically stable.

Notice that the prime mover with the speed governor is described by the following control system

$$\dot{y} = Hy + b\mu\,, \qquad \zeta = q^* y - \beta\mu$$

which can be coupled with our system by taking

$$P_{mec} = P_{mec_0} + \zeta\,, \qquad \mu = s$$

The Domain of Admissible Manoeuvres

As we have already seen a manoeuvre is mathematically described as the passage from some equilibrium to another one and in order to ensure that the manoeuvre is successful, the former equilibrium must belong to the basin of attraction of the new one. Therefore, to find the admissibility of a manoeuvre means to get information about the domain of attraction of an asymptotically stable equilibrium.

Corollary 3.1 *Under assumptions of Theorem 3.2, let the point* $(0, -A^{-1}c(\hat{\delta}), \hat{\delta}, 0)$ *be an asymptotically stable equilibrium; assume that there exists an interval* (δ_1, δ_2) *such that* $\hat{\delta} \in (\delta_1, \delta_2)$ *and* $\varphi(\delta) \neq 0$ *for* $\delta \in (\delta_1, \delta_2)$, $\delta \neq \hat{\delta}$, *and moreover*

$$\int_{\delta_1}^{\delta_2} \varphi(\theta)d\theta \leq 0.$$

Then the set

$$\{0, -A^{-1}c(\delta),\ \tilde{\delta} < \delta < \delta_2,\ \int_{\tilde{\delta}}^{\delta_2} \varphi(\theta)d\theta = 0\}$$

is contained in the domain of attraction of $(0, -A^{-1}c(\hat{\delta}), \hat{\delta}, 0)$.

Proof Consider the set

$$\mathcal{M} = \Big\{s, x, \delta, y\ :\ \tilde{V}(s, x, \delta, y) - V(0, -A^{-1}c(\hat{\delta}), \hat{\delta}) <$$

$$< \frac{1}{2}\int_{\tilde{\delta}}^{\delta_2} \varphi(\theta)d\theta,\ \ \tilde{\delta} < \delta < \delta_2\Big\}.$$

This set is invariant; it is clear that from $\frac{d\tilde{V}}{dt} \leq 0$ it follows that if $(s(0), x(0), \delta(0), y(0)) \in \mathcal{M}$, then

$$\tilde{V}(s(t), x(t), \delta(t), y(t)) - V(0, -A^{-1}c(\hat{\delta}), \hat{\delta}) < \int_{\tilde{\delta}}^{\delta_2} \varphi(\theta)d\theta$$

We see that

$$\tilde{V}(s, x, \delta, y) = y^*Gy + \frac{1}{2}\omega s^2 + \frac{1}{2}[x^* + c^*(\delta)A^{-1}]A[x+$$

$$+A^{-1}c(\delta)] + \frac{1}{2}\chi(\delta) - \frac{1}{2}c^*(\delta)A^{-1}c(\delta)$$

and

$$\int_{\hat\delta}^{\delta}\varphi(\theta)d\theta = \int_{\hat\delta}^{\delta}[\chi'(\theta) - c^*(\theta)A^{-1}c'(\theta) - (c'(\theta))^*A^{-1}c(\theta)]d\theta =$$

$$= \int_{\hat\delta}^{\delta}\frac{d}{d\theta}[\chi(\theta) - c^*(\theta)A^{-1}c(\theta)]d\theta =$$

$$= \chi(\delta) - c^*(\delta)A^{-1}c(\delta) - \chi(\hat\delta) + c^*(\hat\delta)A^{-1}c(\hat\delta).$$

We deduce that

$$\tilde V(s(t), x(t), \delta(t), y(t)) - V(0, -A^{-1}c(\hat\delta), \hat\delta) =$$

$$= y^*(t)Gy(t) + \frac{1}{2}\omega[s(t)]^2 + \frac{1}{2}[x^*(t) + c^*(\delta(t))A^{-1}]A[x(t)+$$

$$+A^{-1}c(\delta(t))] + \frac{1}{2}\int_{\hat\delta}^{\delta(t)}\varphi(\theta)d\theta + \frac{1}{2}[\chi(\hat\delta) - c^*(\hat\delta)A^{-1}c(\hat\delta)]-$$

$$-V(0, -A^{-1}c(\hat\delta), \hat\delta).$$

On the other hand

$$V(0, -A^{-1}c(\hat\delta), \hat\delta) = \frac{1}{2}\chi(\hat\delta) - \frac{1}{2}c^*(\hat\delta)A^{-1}c(\hat\delta)$$

hence

$$y^*(t)Gy(t) + \frac{1}{2}\omega[s(t)]^2 + \frac{1}{2}[x(t) + A^{-1}c(\delta(t))]^*A[x(t)+$$

$$+A^{-1}c(\delta(t))] + \frac{1}{2}\int_{\hat\delta}^{\delta(t)}\varphi(\theta)d\theta < \frac{1}{2}\int_{\hat\delta}^{\delta_2}\varphi(\theta)d\theta$$

that is

$$0 \le y^*(t)Gy(t) + \frac{1}{2}\omega[s(t)]^2 + \frac{1}{2}[x(t) + A^{-1}c(\delta(t))]^*A[x(t)+$$

$$+A^{-1}c(\delta(t))] < \frac{1}{2}\int_{\delta(t)}^{\delta_2}\varphi(\theta)d\theta.$$

It follows that $\int_{\delta(t)}^{\delta_2} \varphi(\theta)d\theta > 0$, hence $\tilde{\delta} < \delta(t) < \delta_2$. The same inequality shows that all solutions in $\mathcal{M}$ are bounded: for $\delta(\cdot)$ this follows from the definition, for x and s from the inequality and for y from the fact that H is a Hurwitz matrix.

According to Barbašin - Krasovskii - La Salle theorem (Theorem 2.5), all solutions in $\mathcal{M}$ have to tend to the maximal invariant set contained in the set where $\frac{d\tilde{V}}{dt} = 0$, hence to the set of equilibria. But under our assumptions, the only equilibrium in $\mathcal{M}$ is $(0, -A^{-1}c(\hat{\delta}), \hat{\delta}, 0)$, hence we have proved that the invariant set $\mathcal{M}$ lies in the domain of attraction of this equilibrium and the manoeuvre of passing from an equilibrium in $\mathcal{M}$ to the given equilibrium associated to $\hat{\delta}$ is admissible.

Let us finally note that the definition of $\tilde{\delta}$ leads to

$$\int_{\hat{\delta}}^{\delta_2} \varphi(\theta)d\theta = \int_{\hat{\delta}}^{\tilde{\delta}} \varphi(\theta)d\theta$$

which is nothing else but the "equal area" criterion well - known by engineers (e.g. Crary, 1947).

The Reduction Principle and Global behaviour of the Synchronous Generator with Prime Mover and Speed Governor

We will make use of the result due to G.A.Leonov (1974) in order to deduce a global behaviour of all solutions of system (3.1.18) corresponding to a generator with a prime mover and speed governor.

Theorem 3.3 *Consider the system* (3.1.18), *assuming that* $A > 0$, $B > 0$, (H, b) *is controllable, and* (q^*, H) *is observable. Moreover, we assume that there exist* $\lambda > 0$, $\epsilon > 0$ *such that:*

a) $\lambda < \lambda_{min}(AB)$, *where* $\lambda_{min}(AB)$ *is the minimal eigenvalue of the matrix* AB*;*

b) $\lambda I + H$ *is a Hurwitz matrix;*

c) $\alpha+\beta-(\lambda+\epsilon)-\Re q^*[(-\lambda+\imath\mu)I-H]^{-1}b \geq 0$ *for all* $\mu \geq 0$;

d) all solutions of the equation

$$\ddot{\theta}+2\sqrt{\lambda\epsilon}\dot{\theta}+\psi(\theta)=0 \tag{3.1.20}$$

are bounded; here $\psi(\theta)=\frac{\omega}{2}\varphi(\theta)$ *is assumed to have exactly two zeros on a period and* $[\psi(\theta)]^2+[\psi'(\theta)]^2 \neq 0$ *for all* θ.

Then every solution tends to an equilibrium of the form $(0,\hat{x},\hat{\delta},0)$.

Proof The frequency domain condition c) gives, by Yakubovich – Kalman – Popov lemma, the existence of γ, w, G such that

$$\omega[\alpha+\beta-(\lambda+\epsilon)]=\gamma^2$$

$$-\frac{\omega q}{2}-Gb=\gamma w$$

$$(\lambda I+H)^*G+G(\lambda I+H)+ww^*=0$$

and since $\lambda I+H$ is a Hurwitz matrix, we have $G \geq 0$. We define

$$W(s,x,\delta,y)=\frac{1}{2}\omega^2 s^2+\frac{1}{2}\omega[x+A^{-1}c(\delta)]^*A[x+A^{-1}c(\delta)]+$$
$$+\omega y^*Gy.$$

Note that

$$W(s,x,\delta,y)=\omega\tilde{V}(s,x,\delta,y)-\frac{\omega}{2}\int_{\hat{\delta}}^{\delta}\varphi(\theta)d\theta-\omega V(0,\hat{x},\hat{\delta}).$$

We deduce that

$$\frac{dW}{dt}(s,x,\delta,y)=\omega\left\{\frac{\partial V}{\partial s}(s,x,\delta)\left[-\frac{\partial V}{\partial \delta}(s,x,\delta)-(\alpha+\beta)s+\right.\right.$$

$$\left.+q^*y\right]+\frac{\partial V}{\partial x}(s,x,\delta)\left[-B\left(\frac{\partial V}{\partial x}(s,x,\delta)\right)^*\right]+\frac{\partial V}{\partial \delta}(s,x,\delta)\cdot$$

$$\left.\cdot\frac{\partial V}{\partial s}(s,x,\delta)\right\}+\omega y^*G(Hy+bs)+\omega(y^*H^*+sb^*)Gy-$$

$$-\frac{\omega}{2}\varphi(\delta)\frac{d\delta}{dt} = -\omega^2 s^2(\alpha+\beta) + \omega s y^* G b + \omega s b^* G y + \omega^2 s q^* y +$$

$$+\omega y^*(GH + H^*G)y - \frac{\omega}{2}\varphi(\delta)\frac{d\delta}{dt} - \frac{\partial V}{\partial x}(s,x,\delta)B\left(\frac{\partial V}{\partial x}(s,x,\delta)\right)^* .$$

We have further

$$\frac{dW}{dt}(s,x,\delta,y) + 2\lambda W(s,x,\delta,y) + \epsilon\left(\frac{d\delta}{dt}\right)^2 + \psi(\delta)\frac{d\delta}{dt} =$$

$$= -\omega^2 s^2(\alpha+\beta) + \omega s y^*(-\frac{\omega q}{2} - \gamma w) + \omega s(-\frac{\omega q^*}{2} - \gamma w^*)y +$$

$$+\omega^2 s q^* y - \omega y^* w w^* y - 2\lambda\omega y^* G y + \lambda\omega^2 s^2 +$$

$$+\lambda\omega[x + A^{-1}c(\delta)]^* A[x + A^{-1}c(\delta)] + 2\lambda\omega y^* G y +$$

$$+\epsilon\omega^2 s^2 + [\psi(\delta) - \frac{\omega}{2}\varphi(\delta)]\frac{d\delta}{dt} - \frac{\partial V}{\partial x}(s,x,\delta)B\left(\frac{\partial V}{\partial x}(s,x,\delta)\right)^* .$$

Now we notice that

$$[x + A^{-1}c(\delta)]^* A[x + A^{-1}c(\delta)] = [Ax + c(\delta)]^* A^{-1}[Ax + c(\delta)] =$$

$$= \frac{\partial V}{\partial x}(s,x,\delta)A^{-1}\left(\frac{\partial V}{\partial x}(s,x,\delta)\right)^*$$

and that $\psi(\delta) = \frac{\omega}{2}\varphi(\delta)$. It follows that

$$\frac{dW}{dt}(s,x,\delta,y) + 2\lambda W(s,x,\delta,y) + \epsilon\left(\frac{d\delta}{dt}\right)^2 + \psi(\delta)\frac{d\delta}{dt} =$$

$$= -\omega(\gamma s + w^* y)^2 - \omega\frac{\partial V}{\partial x}(s,x,\delta)(B - \lambda A^{-1})\left(\frac{\partial V}{\partial x}(s,x,\delta)\right)^* \leq 0$$

since $\lambda < \lambda_{min}(AB)$.

We are now in position to apply the general theorem of Leonov (1974) which we will prove in the Appendix. We have to use the assumption that ψ has exactly two zeros on an interval of the length equal to a period (this happens always when $x_d = x_q$ and for P_{mec} sufficiently large otherwise) and that

$$[\psi(\theta)]^2 + [\psi'(\theta)]^2 \neq 0$$

The theorem of Leonov gives boundedness of $\delta(t)$ for every solution of (3.1.18).

Note further that

$$\dot{x}(t) = -B[Ax(t) + c(\delta(t))].$$

and since $-(BA)^*A + A(-BA) = -2ABA < 0$ it follows from the Liapunov theorem that $-BA$ is a Hurwitz matrix (in fact we already know that all its eigenvalues are real and negative but it is nice to use once more the Liapunov theorem!) and from boundedness of δ we deduce boundedness of x.

Let now

$$\nu > \frac{1}{8\lambda\epsilon} \sup_{\delta} \psi^2(\delta)$$

$$U(t) = W(s(t), x(t), \delta(t), y(t)) - \nu.$$

We have

$$\dot{U}(t) + 2\lambda U(t) + \epsilon\omega^2 s^2(t) + \omega s(t)\psi(\delta(t)) + 2\lambda\nu =$$

$$= \frac{dW}{dt}(s(t), x(t), \delta(t), y(t)) + 2\lambda W(s(t), x(t), \delta(t), y(t)) +$$

$$+\epsilon\left(\frac{d\delta(t)}{dt}\right)^2 + \psi(\delta(t))\frac{d\delta(t)}{dt} \leq 0.$$

In addition,

$$\epsilon\omega^2 s^2(t) + \omega s(t)\psi(\delta(t)) + 2\lambda\nu =$$

$$= \epsilon\left[\omega s(t) + \frac{1}{2\epsilon}\psi(\delta(t))\right]^2 + 2\lambda\nu - \frac{1}{4\epsilon}[\psi(\delta(t))]^2 > 0.$$

for all t, hence

$$\frac{d}{dt}(e^{2\lambda t}U(t)) \leq 0, \qquad U(t)e^{2\lambda t} \leq U(0),$$

$$W(s(t),x(t),\delta(t),y(t)) \leq \nu + e^{-2\lambda t}[W(s(0),x(0),\delta(0),y(0)) - \nu].$$

Inspecting the formula for W and knowing that $x(\cdot)$ and $\delta(\cdot)$ are bounded for $t > 0$, we deduce that $s(\cdot)$ is bounded for $t > 0$. On the other hand, $\lambda I + H$ is a Hurwitz matrix, hence H is a Hurwitz matrix and from $\dot{y}(t) = Hy(t) + bs(t)$ we finally obtain that $y(\cdot)$ is bounded and therefore we have proved that all solutions of the system are bounded.

As in the proof of the previous theorem, we see that

$$\frac{d\tilde{V}}{dt}(s,x,\delta,y) = -\frac{\partial V}{\partial x}(s,x,\delta)B\left(\frac{\partial V}{\partial x}(s,x,\delta)\right)^* - (\lambda+\epsilon)\omega s^2 -$$

$$-2\lambda y^*Gy - \frac{1}{\omega}(\gamma s + w^*y)^2.$$

From the Barbašin – Krasovskii – La Salle theorem we deduce that every solution tends to the largest invariant set contained in the set where

$$s = 0, \qquad \frac{\partial V}{\partial x}(s,x,\delta) = 0, \qquad y^*Gy = 0$$

We deduce that for every solution we must have $s(t) \to 0$, $y(t) \to 0$ (H is a Hurwitz matrix), $Ax(t) + c(\delta(t)) \to 0$, $\tilde{V}(s(t),x(t),\delta(t),y(t)) \to \tilde{V}_\infty$. Since $y^*(t)Gy(t) \to 0$ we deduce that $V(s(t),x(t),\delta(t)) \to \tilde{V}_\infty$. On the other hand, we can write

$$V(s(t),x(t),\delta(t)) = \frac{1}{2}\frac{\partial V}{\partial x}(s(t),x(t),\delta(t))A^{-1}\cdot$$

$$\cdot\left[\frac{\partial V}{\partial x}(s(t),x(t),\delta(t))\right]^* + \frac{1}{2}\omega s^2(t) + \Theta(\delta(t))$$

hence $\lim_{t\to\infty}\Theta(\delta(t)) = \tilde{V}_\infty$.

Here $\Theta(\cdot)$ is a non-constant analytic function. We will check that $\lim_{t\to\infty} \delta(t)$ exists. If it does not, then there are $\delta' < \delta''$ and $t_j' \to \infty$, $t_j'' \to \infty$ such that $\delta(t_j') \to \delta'$, $\delta(t_j'') \to \delta''$. Let $\delta' < \bar{\delta} < \delta''$; for j large enough we have $\delta(t_j') < \bar{\delta} < \delta(t_j'')$; hence there is $\bar{t}_j \in (t_j', t_j'')$ such that $\delta(\bar{t}_j) = \bar{\delta}$. On the other hand, $\bar{t}_j \to \infty$ hence $\Theta(\delta(\bar{t}_j)) \to \tilde{V}_\infty$ and we obtain that $\Theta(\bar{\delta}) = \tilde{V}_\infty$ for every $\bar{\delta} \in [\delta', \delta'']$ which is a contradiction. We have thus proved that $s(t) \to 0$, $y(t) \to 0$, $\delta(t) \to \hat{\delta}$ and since $Ax(t) + c(\delta(t)) \to 0$, we have $x(t) \to -A^{-1}c(\hat{\delta})$, i.e., every solution tends to an equilibrium as claimed.

3.2 Stabilization of Class of Steam Turbines for Heat-Electricity Generation

The steam turbines with regulated bleedings are applied in combined heat-electricity generation. The electric energy is obtained from the rotation mechanical energy of the turbine, while the thermal energy is obtained from the energy of the steam which is extracted at the regulated bleeding and supplied to the thermal consumer at some constant parameters.

The constant frequency of the electric energy supply is ensured by rotating speed control and the constant parameters of the thermal energy supply are ensured by bleeding parameters (pressure, flow) control. It follows that at least two kinds of disturbances can occur; electrical load disturbances and thermal load disturbances. If short–period disturbances are considered, obviously a stability problem with respect to initial conditions occurs.

Moreover, the operating conditions of steam turbines make impossible manual control of parameters and manual changes of the operating steady–state (the manual "manoeuvres"). The automatic feedback control system of the turbine, with its feedback connections can introduce new physical phenomena which can be viewed as instability. For such reasons the stability analysis for the system turbine–feedback controller is very important in the design and manufacturing of high–performance power–generating machines.

Usually the mathematical models of steam turbine dynamics are linearized around a steady–state (V.A.Ivanov 1971, 1982, IEEE Committee Report on *Dynamic Models for Steam and Hydro Turbines in Power System Studies*, 1973). As it is well known from first approximation stability (e.g. Theorem 2.1 of this monograph) such models can give at most local stability. The global stability requirements can be analyzed only by using global models. But the global models for thermal processes are highly nonlinear. Under some assumptions bilinear models can be used, reproducing almost all basic properties of the physical phenomena which are studied (Vl.Răsvan, 1981).

Configurations and Mathematical Models

In the following, we will consider small power steam turbines (1 – 12 MW), without reheating and single regulating bleeding. It will be assumed that the steam volumes enclosed in the turbine cylinders are small enough as to allow neglecting of the time constants introduced by these volumes. Consequently, the following model of one turbine as a controlled dynamical plant is obtained:

$$T_a\dot{s} = \alpha\mu_1 + (1-\alpha)\mu_2\Pi_s - \nu_g$$

$$T_p\dot{\Pi}_s = \mu_1 - \beta_1\mu_2\Pi_s - \beta_2 g_p \tag{3.2.1}$$

where s is deviation of the rotating speed with respect to the imposed value, Π_s is the bled–steam pressure, ν_g – the mechanical load (the power supplied to the driven synchronous generator) and g_p – the thermal load – the bled–steam flow. The variables μ_1 and μ_2 represent the control signals – the positions of the control valves – subject to the following restrictions

$$0 \le \mu_i \le 1 \qquad (i = 1, 2).$$

It is worth to mentioning that all variables are scaled per unit i.e., reported to some rated values. If the distance between the steam bleeding and the consumer is quite large, the propagation effects along the steam pipes have to be taken into account. However, if some additional assumptions are introduced, one of the following lumped-parameter models (i.e., described by ordinary differential equations) can be used:

$$\begin{aligned} T_a\dot{s} &= \alpha\mu_1 + (1-\alpha)\mu_2\Pi_s - \nu_g \\ T_p\dot{\Pi}_s &= \mu_1 - (\beta_1\mu_2 + \beta_2\alpha_p)\Pi_s + \beta_2\alpha_p\xi_p \\ T_c\dot{\xi}_p &= \alpha_p\Pi_s - (\alpha_p + \psi_s)\xi_p \end{aligned} \tag{3.2.2}$$

or

$$\begin{aligned} T_a\dot{s} &= \alpha\mu_1 + (1-\alpha)\mu_2\Pi_s - \nu_g \\ T_p\dot{\Pi}_s &= \mu_1 - (\beta_1\mu_2 + \beta_2\frac{\alpha_p\psi_s}{\alpha_p + \psi_s})\Pi_s \end{aligned} \tag{3.2.3}$$

The Invariant Set of the Models

When the mathematical model of some physical process is considered, its properties have to be deduced in a rigorous way from the assumptions and no additional physical arguments can be introduced after the model has been adopted (additional physical assumptions to an already constituted mathematical model represents in fact a change of the model).

The first property that has to be verified is the existence of solutions. For the models (3.2.2) and (3.2.3) the RHS of the ordinary differential equations is such that the existence and uniqueness conditions are fulfilled. Therefore, given μ_1, μ_2, and the initial conditions, a unique solution always exists.

A second property of the models is established starting from the physical significance of some state variables: $\Pi_s(t)$, $\xi_p(t)$ represent steam pressures and they have to be nonnegative. Mathematically speaking this means that the models must have an invariant set: nonnegative initial conditions $\Pi_s(0) \geq 0$, $\xi_p(0) \geq 0$ must imply nonnegativeness along the trajectory, i.e., $\Pi_s(t) \geq 0$, $\xi_p(t) \geq 0$ for all $t \geq 0$. Consequently it is necessary to prove

Theorem 3.4 *a) If $\Pi_s(0) \geq 0$ then the component $\Pi_s(t)$ of the solution of (3.2.3) satisfies $\Pi_s(t) \geq 0$ for all $t \geq 0$.*

b) If $\Pi_s(0) \geq 0$, $\xi_p(0) \geq 0$ then the components $\Pi_s(t)$, $\xi_p(t)$ of the solution of (3.2.2) satisfy $\Pi_s(t) \geq 0$, $\xi_p(t) \geq 0$ for all $t \geq 0$.

Proof a) By writing the representation of the solution with μ_1, μ_2 being functions of time, we find that:

$$\Pi_s(t) = \Pi_s(0) \exp\left[-\frac{1}{T_p}\int_0^t \left(\beta_2 \frac{\alpha_p \psi_s}{\alpha_p + \psi_s} + \beta_1 \mu_2(\tau)\right) d\tau\right] +$$

$$+\frac{1}{T_p}\int_0^t \mu_1(\tau) \exp\left[-\frac{1}{T_p}\int_\tau^t \left(\beta_2 \frac{\alpha_p \psi_s}{\alpha_p + \psi_s} + \beta_1 \mu_2(\theta)\right) d\theta\right] d\tau .$$

It is obvious that due to the fact that $\mu_1(\tau) \geq 0$, if $\Pi_s(0) \geq 0$, $\Pi_s(t) \geq 0$ for all $t \geq 0$.

b) By writing (3.2.2) in the integral form, it follows

$$\Pi_s(t) = \Pi_s(0) \exp\left(-\frac{1}{T_p}\int_0^t (\beta_1 \mu_2(\tau) + \beta_2 \alpha_p) d\tau\right) +$$

$$+\frac{1}{T_p}\int_0^t (\mu_1(\tau) + \beta_2 \alpha_p \xi_p(\tau)) \exp\left(-\frac{1}{T_p}\int_\tau^t (\beta_1 \mu_2(\theta) + \beta_2 \alpha_p) d\theta\right) d\tau$$

$$\xi_p(t) = \xi_p(0) \exp\left(-\frac{1}{T_c}(\alpha_p + \psi_s) t\right) +$$

$$+\frac{\alpha_p}{T_c}\int_0^t \Pi_s(\tau) \exp\left(-\frac{1}{T_c}(\alpha_p + \psi_s)(t - \tau)\right) d\tau .$$

Assume first that $\Pi_s(0) > 0$, $\xi_p(0) > 0$. Define the set $\mathcal{M} = \{t, \Pi_s(\tau) > 0, \xi_p(\tau) > 0, 0 \leq \tau < t\}$. This set is nonempty and let $\theta =$

$\sup \mathcal{M}$. If (T_1, T_2) is the existence interval, assume $\theta < T_2$ (otherwise the theorem is proved). Because $\Pi_s(\tau) > 0, \xi_p(\tau) > 0, 0 \leq \tau < \theta$ and $\mu_1(\tau) > 0$, for all τ, it follows that $\Pi_s(\theta) > 0$ and $\xi_p(\theta) > 0$. But the continuity will imply also that $\Pi_s(\theta + \eta) > 0$, $\xi_p(\theta + \eta) > 0$ for $\eta > 0$ sufficiently small. This contradicts the maximalness of θ hence $\Pi_s(t) > 0$, $\xi_p(t) > 0$ on the whole existence interval.

Assume now that $\Pi_s(0) \geq 0$, $\xi_p(0) \geq 0$ and take some $\epsilon > 0$. Denote by $\Pi_s^\epsilon(t)$, $\xi_p^\epsilon(t)$ the solutions of (3.2.2) corresponding to the initial conditions $\Pi_s(0) + \epsilon$, $\xi_p(0) + \epsilon$. Obviously $\Pi_s^\epsilon(t) > 0$, $\xi_p^\epsilon(t) > 0$ on the existence interval. Let now $\epsilon \to 0$. From the continuous dependence theorem it follows that $\Pi_s^\epsilon(t) \to \Pi_s(t)$, $\xi_p^\epsilon(t) \to \xi_p(t)$ uniformly on compact sets and $\Pi_s(t) \geq 0$, $\xi_p(t) \geq 0$ what ends the proof.

Steady–States

It has been shown already in the introductory part that current operation of the steam turbine requires maintaining of some steady–state and passing from one steady–state to another (manoeuvres). The steady–state equations are given by

$$\begin{aligned}
&\alpha\mu_1^0 + (1-\alpha)\mu_2^0\Pi_s^0 = \mu_g \\
&\mu_1^0 - (\beta_1\mu_2^0 + \beta_2\alpha_p)\Pi_s^0 + \beta_2\alpha_p\xi_p^0 = 0 \qquad (3.2.2') \\
&\alpha_p\Pi_s^0 = (\alpha_p + \psi_s)\xi_p^0
\end{aligned}$$

for system (3.2.2) and by

$$\begin{aligned}
&\alpha\mu_1^0 + (1-\alpha)\mu_2^0\Pi_s^0 = \nu_g \\
&\mu_1^0 - \left(\beta_1\mu_2^0 + \beta_2\frac{\alpha_p\psi_s}{\alpha_p + \psi_s}\right)\Pi_s^0 = 0 \qquad (3.2.3')
\end{aligned}$$

for system (3.2.3).

Consider first system (3.2.2'): there are 3 equalities and 4 unknowns. The mechanical power ν_g – in fact the electrical power required from the power generating unit – is imposed. One can impose still one variable for the thermal consumer or for the bleeding (Π_s^0 – the bled–steam pressure, ξ_p^0 – the steam pressure at the consumer) and in this case all unknowns can be found. Finding μ_1^0 and μ_2^0 means in fact to find the references of rotating speed and bleeding pressure controllers. In practice, for imposed ν_g (the mechanical load) and Π_s^0 (or ξ_p^0) (the thermal load) the references can be found allowing for realization of the desired steady–state. By modifying μ_1^0 and μ_2^0 the steady–state is modified i.e., manoeuvres are performed. The determination of μ_1^0 and μ_2^0 for imposed ν_g and Π_s^0 (or ξ_p^0) corresponds to the so called *electric load schedule operation* which is met in the case of condensing turbines.

For the case of back–pressure turbines the operation is different: the thermal consumer down stream cannot accept a unlimited quantity of steam. Consequently, the steam flow entering the low pressure cylinder of turbine (after the regulated bleeding) is also imposed i.e., μ_2^0 – which is proportional to the steam flow – is also fixed. For the compatibility of system (3.2.2') it is necessary to deduce ν_g – the available electric power – from the compatibility condition which ensures the existence of the steady–state.

This operation is called *thermal load schedule.* Both types of operation schedules are well–known to practitioners.

In the same manner, if the system (3.2.3') is considered, ν_g and Π_s^0 are imposed for the case of electric load schedule and μ_2^0 and Π_s^0 are imposed for the case of thermal load schedule.

Two problems still occur in steady–state analysis. The first one concerns compatibility and the number of solutions for the nonlinear (in fact bilinear) systems of algebraic equations defining the steady–states.

In the case of system (3.2.2'), the knowledge of one of the bled–steam parameters allows a unique determination of all other parameters: if Π_s^0

(or ξ_p^0) is known, the system becomes linear. In the case of (3.2.3'), if Π_s^0 is known, a linear system with respect to μ_1^0, μ_2^0 is obtained. In both cases the determinant is given by

$$\Delta = \begin{vmatrix} \alpha & 1-\alpha \\ 1 & -\beta_1 \end{vmatrix} = -(1-\alpha+\alpha\beta_1) \neq 0$$

because $0 < \alpha < 1$, $0 < \beta_1 < 1$. If μ_2^0 is imposed, μ_1^0 and ν_g are obtained immediately. In all cases a unique solution can be found.

The second problem concerns admissibility of the steady–states. It is obvious that one must always obtain $0 < \mu_1^0 < 1$ and $0 < \nu_g < 1$ (for the case of the thermal load schedule). The fulfilment of these inequalities is achievable if taking into account the specific properties of each turbine $(\alpha, \beta_1, \beta_2, \alpha_p, \psi_s)$, the values of Π_s^0, ξ_p^0, ν_g, μ_2^0 are prescribed accordingly. In practice this goal is achieved by using some graphical representations of the steady–state equations (3.2.2'), (3.2.3') – the so–called steam consumption diagrams of the turbine – which are supplied by the manufacturer.

In the following it will be assumed that the steady–state is always admissible.

It is worth mentioning that the steady–state of the variable s does not result from the equations but it is imposed $(s = 0)$ by the synchronism with the grid.

Systems in Deviations and Inherent Stability

Denoting

$$x_1 = \Pi_s - \Pi_s^0, \qquad u_i = \mu_i - \mu_i^0 \quad (i = 1, 2)$$

$$\zeta_p = \xi_p - \xi_p^0$$

where Π_s^0, μ_i^0, ξ_p^0 correspond to some steady–state, the following systems in deviations are obtained:

$$T_a\dot{s} = (1-\alpha)\mu_2^0 x_1 + \alpha u_1 + (1-\alpha)(\Pi_s^0 + x_1)u_2$$

$$T_p\dot{x}_1 = -(\beta_1\mu_2^0 + \beta_2\alpha_p)x_1 + \beta_2\alpha_p\zeta_p + u_1 -$$

$$-\beta_1(\Pi_s^0 + x_1)u_2 \qquad (3.2.4)$$

$$T_c\dot{\zeta}_p = \alpha_p x_1 - (\alpha_p + \psi_s)\zeta_p$$

for the case of (3.2.2), and

$$T_a\dot{s} = (1-\alpha)\mu_2^0 x_1 + \alpha u_1 + (1-\alpha)(\Pi_s^0 + x_1)u_2$$

$$T_p\dot{x}_1 = -\left(\beta_1\mu_2^0 + \beta_2\frac{\alpha_p\psi_s}{\alpha_p + \psi_s}\right)x_1 + u_1 - \beta_1(\Pi_s^0 + x_1)u_2 \qquad (3.2.5)$$

for the case (3.2.3).

In this way stability of the steady–state is reduced to stability of the zero solution corresponding to $u_1 = u_2 = 0$. When the control functions are identically zero, the system in deviations becomes linear and the stability problem is a problem of inherent stability which can be studied using the Hurwitz criterion. The characteristic equation of system (3.2.4) is

$$\det\begin{pmatrix} T_a\lambda & -(1-\alpha)\mu_2^0 & 0 \\ 0 & T_p\lambda + \beta_1\mu_2^0 + \beta_2\alpha_p & -\beta_2\alpha_p \\ 0 & -\alpha_p & T_c\lambda + \alpha_p + \psi_s \end{pmatrix} =$$

$$T_a\lambda[(T_p\lambda + \beta_1\mu_2^0 + \beta_2\alpha_p)(T_c\lambda + \alpha_p + \psi_s) - \beta_2\alpha_p^2] = 0$$

It is obvious that the 2–nd degree polynomial between the braces has its roots with negative real parts, due to its positive coefficients; however the zero root endows system (3.2.4) only with simple stability not with asymptotic stability.

The characteristic equation of (3.2.5) is

$$\det \begin{pmatrix} T_a\lambda & -(1-\alpha)\mu_2^0 \\ 0 & T_p\lambda + \beta_1\mu_2^0 + \beta_2\frac{\alpha_p\psi_s}{\alpha_p+\psi_s} \end{pmatrix} = 0$$

and it has a negative real root and a zero root; therefore system (3.2.5) is also only stable, not asymptotically stable.

Both models have only simple inherent stability while in practice asymptotic stability is required. A stabilization is thus necessary; it is accomplished using the deviations of the control functions. These deviations are constructed as state feedback corrections. The feedback structure is also imposed by other considerations (robustness to uncertainties, disturbance rejection) which will not be tackled here.

Stabilization

The stabilization structures for steam turbines are known since they entered in exploitation. For turbines with regulated bleedings it has been recognized relatively early that the solution which uses separate regulation for rotating speed and pressure at bleeding is not satisfactory even from the stability view-point. It is why the following structure was considered:

$$u_1 = -k_{11}s - k_{12}x_1$$

$$u_2 = -k_{21}s + k_{22}x_1\,, \quad k_{ij} > 0. \tag{3.2.6}$$

Such structure is obtained when analyzing the currently used linearized models for machines (M.Tolle, 1906; I.N.Voznesenskii, 1938; I.I.Kirillov, 1988; V.A.Ivanov, 1971, 1982).

Consider the model (3.2.5), one of the most widely used in the theory of regulation for steam turbines and take u_i in the form given by (3.2.6) with higher order terms neglected. The matrix for the linear system has the characteristic equation

$$\det\begin{pmatrix} T_a\lambda + \alpha k_{11} + & -(1-\alpha)\mu_2^0 + \alpha k_{12} - \\ +(1-\alpha)\Pi_s^0 k_2, & -(1-\alpha)\Pi_s^0 k_{22} \\ & \\ k_{11} - \beta_1\Pi_s^0 k_{21}, & T_p\lambda + \beta_1\mu_2^0 + \beta_2\dfrac{\alpha_p\psi_s}{\alpha_p+\psi_s} + \\ & +k_{12} + \beta_1\Pi_s^0 k_{22} \end{pmatrix} = 0$$

A direct check shows that the coefficients of the 2–nd order characteristic polynomial are positive for $k_{ij} > 0$, hence the structure considered here leads to local stability.

A similar analysis gives the same result for the model with three state variables. For practical purposes it is important to have global stability. For bilinear systems, and our system is a bilinear one, global stabilization was studied by M.Slemrod (1978) and J.P.Quinn (1980). For the system

$$\dot{x} = Ax + Bxu\,, \quad x \in R^n\,,\ u \in R$$

it was assumed that: i) there exists $Q > 0$ such that $QA + A^*Q = 0$; ii) $x_0^* e^{A^*t} QBe^{At} x_0 \equiv 0$ implies $x_0 = 0$. Under such assumptions the choice

$$u = -x^*QBx$$

leads to global stability; the stability is proved by using the Liapunov function

$$V(x) = \frac{1}{2}x^*Qx$$

If we have control constraints like $|u| \leq 1$ the control is taken "with saturation" as

$$u(t) = \begin{cases} -1 & \text{if } x^*(t)QBx(t) > 1 \\ -x^*(t)QBx(t) & \text{if } |x^*(t)QBx(t)| \leq 1 \\ 1 & \text{if } x^*(t)QBx(t) < -1 \end{cases}$$

The controls "with saturation" are well – known in turbine regulation. What is less – known however, is the feedback control quadratic with respect to the state variables. In what follows, we will apply the Liapunov function method to our models, taking into account their particular properties, mainly the existence of an invariant set.

We will prove our results in a more general setting, for systems of the form

$$\dot{x} = A_{11}x + A_{12}y + \sum_{j=1}^{k} a_j^0 \mu^j + \sum_{j=k+1}^{r} a_j \mu^j y^j + a$$

$$\dot{y} = A_{22}y + \sum_{j=1}^{k} b_j^0 \mu^j + \sum_{j=k+1}^{r} b_j \mu^j y^j + b \qquad (3.2.7)$$

with the control constraints

$$0 \leq \mu_{min}^j \leq \mu^j \leq \mu_{max}^j, \quad j = 1, \ldots, r$$

Proposition 3.1 *Assume that* b_i^0 *and* b *have positive coordinates,* $b_i^j \geq 0$ *for* $i \neq j$ *and the off–diagonal elements of* A_{22} *are positive. Then if* $y(0)$ *has positive coordinates the same is true for* $y(t)$, $t \geq 0$.

Proof We write the equations for y^j in the form

$$\dot{y}^j(t) = \alpha_j(t)y^j(t) + \beta_j(t)$$

where $\alpha_j(t)$ and $\beta_j(t)$ are bounded and $\beta_j(t) \geq 0$ for all t for which $y^k(t) \geq 0$, $k \neq j$. The proof proceeds further in the standard way (see e.g. Theorem 3.4).

We state the stabilization problem in the same way as for the specific situation of the steam turbines. The steady–state (constant solution) has to be defined in order to keep outputs, at given values r, of the form $p_j^* x + q_j^* y$. Thus, we write the system

$$A_{11}x + A_{12}y + \sum_{i=1}^{k} a_i^0 \mu^i + \sum_{i=k+1}^{r} \mu^i a_i y^i + a = 0$$

$$A_{22}y + \sum_{i=1}^{k} b_i^0 \mu^i + \sum_{i=k+1}^{r} \mu^i b_i y^i + b = 0 \tag{3.2.8}$$

$$p_i^* x + q_i^* y = \rho_i^0 , \quad i = 1, \ldots, r$$

If among some outputs which are to be kept constant, we have the variables y^i, $i = k+1, \ldots, r$, then the system to be solved for the remaining state coordinates and for the controls is linear and, in general, it has a unique solution. This was in fact the case for the models of steam turbines.

We assume that the solution x_0, y_0, μ_0^i of (3.2.8) is unique and introduce the deviations

$$\tilde{x} = x - x_0 , \qquad \tilde{y} = y - y_0 , \qquad u_i = \mu_i - \mu_0^i .$$

Denote

$$\tilde{A}_{12} = A_{12} + \sum_{i=k+1}^{r} \mu_0^i A_i , \qquad \tilde{A}_{22} = A_{22} + \sum_{i=k+1}^{r} \mu_0^i B_i$$

where A_i has a_i as i–th column and all other columns are zero; B_i has b_i as i–th column all other columns being zero.

The system for the deviations is

$$\frac{d}{dt}\tilde{x} = A_{11}\tilde{x} + \tilde{A}_{12}\tilde{y} + \sum_{i=1}^{k} a_i^0 u^i + \sum_{i=k+1}^{r} (\tilde{y}^i + y_0^i) a_i u^i$$

$$\frac{d}{dt}\tilde{y} = \tilde{A}_{22}\tilde{y} + \sum_{i=1}^{k} b_i^0 u^i + \sum_{i=k+1}^{r} (\tilde{y}^i + y_0^i) b_i u^i \tag{3.2.9}$$

We will make now the main assumptions

a) $\tilde{A}_{22}$ is a Hurwitz matrix;

b) A_{11} has all eigenvalues on the imaginary axis with simple elementary divisors.

Under these assumptions, for $u^i = 0$, the zero solution of (3.2.9) is stable and we look for feedback control ensuring global asymptotic stability.

Since A_{11} and $\tilde{A}_{22}$ do not have common eigenvalues the matrix equation

$$T\tilde{A}_{22} - A_{11}T = -\tilde{A}_{12}$$

has a unique solution. The same assumptions lead to existence of $P_1 > 0$, $P_2 > 0$ such that

$$P_1A_{11} + A_{11}^*P_1 = 0\,, \qquad P_2\tilde{A}_{22} + \tilde{A}_{22}^*P_2 = -I$$

We can now construct the Liapunov function

$$V(\tilde{x}, \tilde{y}) = (\tilde{x} + T\tilde{y})^*P_1(\tilde{x} + T\tilde{y}) + \tilde{y}^*P_2\tilde{y} \tag{3.2.10}$$

Since $P_1 > 0$, $P_2 > 0$, this function is positively definite, thus there exists $\lambda_0 > 0$ such that

$$V(\tilde{x}, \tilde{y}) \geq \lambda_0 \left(|\tilde{x}|^2 + |\tilde{y}|^2\right)$$

The derivative with respect to (3.2.9) is

$$\frac{dV}{dt} = -\tilde{y}^*\tilde{y} + \sum_{i=1}^{k} u^i l_i(\tilde{x}, \tilde{y}) +$$

$$+ \sum_{i=k+1}^{r} u^i(\tilde{y}^i + y_0^i) l_i(\tilde{x}, \tilde{y}) \tag{3.2.11}$$

where the linear forms $l_i(\tilde{x}, \tilde{y})$, $i = 1, \ldots, r$ are defined as follows:

$$l_i(\tilde{x}, \tilde{y}) = (a_i^0 + Tb_i^0)^*P_1(\tilde{x} + T\tilde{y}) + (\tilde{x} + T\tilde{y})^*P_1(a_i^0 + Tb_i^0) +$$

$$+(b_i^0)^*P_2\tilde{y}+\tilde{y}^*P_2b_i^0\,,\quad i=1,\dots,k,$$

$$l_i(\tilde{x},\tilde{y})=(a_i+Tb_i)^*P_1(\tilde{x}+T\tilde{y})+(\tilde{x}+T\tilde{y})^*P_1(a_i+Tb_i)+$$

$$+(b_i)^*P_2\tilde{y}+\tilde{y}^*P_2b_i\,,\quad i=k+1,\dots,r, \tag{3.2.12}$$

We choose now the control functions as:

$$u^i=\begin{cases}\mu^i_{max}-\mu^i_0\,, & \text{if } -\gamma_i l_i(\tilde{x},\tilde{y})>\mu^i_{max}-\mu^i_0\\ -\gamma_i l_i(\tilde{x},\tilde{y})\,, & \text{if } -\gamma_i l_i(\tilde{x},\tilde{y})\in\left[\mu^i_{min}-\mu^i_0,\mu^i_{max}-\mu^i_0\right]\\ \mu^i_{min}-\mu^i_0\,, & \text{if } -\gamma_i l_i(\tilde{x},\tilde{y})<\mu^i_{min}-\mu^i_0\end{cases}$$

where $\gamma_i>0$ are arbitrary.

For such control functions, by taking into account the fact that the system has an invariant set of interest, where $\tilde{y}^i+y_0^i\geq 0$, we deduce that on this set $\dfrac{dV}{dt}\leq 0$ hence the considered equilibrium is stable. To deduce asymptotic stability we will consider again the theorem of Barbašin – Krasovskii – La Salle. The set where $\dfrac{dV}{dt}=0$ corresponds to

$$\tilde{y}=0,\quad u^i=0,\quad i=1,\dots,k,$$

$$u^i(\tilde{y}^i+y_0^i)=0,\quad i=k+1,\dots,r$$

It follows that for every trajectory contained in this set

$$x(t)=e^{A_{11}t}x_0$$

and

$$(a_i^0+Tb_i^0)^*P_1e^{A_{11}t}x_0\equiv 0\,,$$

$$(a_i+Tb_i)^*P_1e^{A_{11}t}x_0\equiv 0$$

Defining the matrix

$$Q=P_1\left[a_1^0+Tb_1^0\ \dots\ a_k^0+Tb_k^0\ a_{k+1}+Tb_{k+1}\ \dots\ a_r+Tb_r\right]$$

we see that if (Q^*, A_{11}) is observable, then the above identities imply $x_0 = 0$

Under such observability assumption the only invariant set contained in the set where $\frac{dV}{dt} = 0$ is the equilibrium hence this equilibrium is asymptotically stable.

Let us show how the procedure described above works for the specific situation for steam turbine models. Consider the model with two phase coordinates. Here

$$A_{11} = 0, \qquad \tilde{A}_{12} = (1-\alpha)\frac{\mu_2^0}{T_a},$$

$$\tilde{A}_{22} = -\left(\beta_1\mu_2^0 + \beta_2\frac{\alpha_p\psi_s}{\alpha_p+\psi_s}\right)/T_p,$$

$$a_1^0 = \frac{\alpha}{T_a}, \qquad a_2 = \frac{1-\alpha}{T_a},$$

$$b_1^0 = \frac{1}{T_p}, \qquad b_2 = -\frac{\beta_1}{T_p}$$

The matrix T from above, is replaced with scalar defined by

$$T = \frac{T_p}{T_a}\frac{(1-\alpha)\mu_2^0}{\beta_1\mu_2^0 + \beta_2\frac{\alpha_p\psi_s}{\alpha_p+\psi_s}},$$

$P_1 > 0$ is an arbitrary scalar and

$$P_2 = \frac{T_p}{2\left(\beta_1\mu_2^0 + \beta_2\frac{\alpha_p\psi_s}{\alpha_p+\psi_s}\right)}.$$

Since in this case Q is a scalar, the observability condition is automatically satisfied. The linear functions in the definition of controls are

$$l_1(s, x_1) = \left(\alpha + \frac{(1-\alpha)\mu_2^0}{\beta_1\mu_2^0 + \beta_2\frac{\alpha_p\psi_s}{\alpha_p+\psi_s}}\right)\cdot$$

$$\cdot\left(s + \frac{T_p}{T_a}\frac{(1-\alpha)\mu_2^0}{\beta_1\mu_2^0 + \beta_2\frac{\alpha_p\psi_s}{\alpha_p+\psi_s}}x_1\right) + P_0x_1$$

$$l_2(s,x_1) = \frac{(1-\alpha)\beta_2\frac{\alpha_p\psi_s}{\alpha_p+\psi_s}}{\beta_1\mu_2^0+\beta_2\frac{\alpha_p\psi_s}{\alpha_p+\psi_s}}\left(s+\frac{T_p}{T_a}\frac{(1-\alpha)\mu_2^0}{\beta_1\mu_2^0+\beta_2\frac{\alpha_p\psi_s}{\alpha_p+\psi_s}}x_1\right) -$$

$$-\beta_1 P_0 x_1$$

where $P_0 = 1/P_1 > 0$ is arbitrary. This control functions have the structure considered in the linear theory. In our construction we have the freedom in choosing γ_1, γ_2, P_0 just little smaller than in the general one corresponding to arbitrary $k_{ij} > 0$. A similar analysis may be performed for the system with three state variables.

Let us note now that in our situation we have in fact exponential stability. To this end, let us look at the linear part of the system with feedback control; it reads

$$\frac{d\tilde{x}}{dt} = A_{11}\tilde{x} + \tilde{A}_{12}\tilde{y} - \sum_{j=1}^{k}\gamma_j a_j^0 l_j(\tilde{x},\tilde{y}) - \sum_{j=k+1}^{r}\gamma_j a_j y_0^j l_j(\tilde{x},\tilde{y})$$

$$\frac{d\tilde{y}}{dt} = \tilde{A}_{22}\tilde{y} - \sum_{j=1}^{k}\gamma_j b_j^0 l_j(\tilde{x},\tilde{y}) - \sum_{j=k+1}^{r}\gamma_j b_j y_0^j l_j(\tilde{x},\tilde{y})$$

Considering again the Liapunov function (3.2.10), for the derivative with respect to the above linearized system, we obtain

$$\frac{dV}{dt} = -\tilde{y}^*\tilde{y} - \sum_{j=1}^{k}\gamma_j[l_j(\tilde{x},\tilde{y})]^2 - \sum_{j=k+1}^{r}\gamma_j y_0^j[l_j(\tilde{x},\tilde{y})]^2 \le 0$$

Using the Barbašin – Krasovskii – La Salle argument again, we find that under the observability assumption, we have asymptotic stability, and hence exponential stability in the neighbourhood of the origin. On the other hand, the global asymptotic stability means (e.g. A.Halanay, 1966)

$$|\tilde{x}(t)| + |\tilde{y}(t)| \le \psi(t)\chi(|\tilde{x}(0)| + |\tilde{y}(0)|)$$

with $\chi(\rho)$ increasing and $\psi(t) \to 0$ for $t \to \infty$.

Let K be a compact interval,

$$\beta_1 = \max_{\rho \geq \delta_0, \rho \in K} \frac{\chi(\rho)}{\rho}, \qquad \tilde{\psi}(t) = \max\{\psi(t), e^{-\alpha t}\},$$

$$\tilde{\beta} = \max\{\beta_0, \beta_1\}$$

Here $(\delta_0, \beta_0, \alpha)$ are the parameters corresponding to the local exponential stability. Since all solutions are bounded, they do not leave a compact and we have the global estimate

$$|\tilde{x}(t)| + |\tilde{y}(t)| \leq \tilde{\beta}\,(|\tilde{x}(0)| + |\tilde{y}(0)|)\,\tilde{\psi}(t)$$

Such estimate gives the exponential stability (A.Halanay, 1966).

Appendix 1

The Theorem of G.A. Leonov

This result which is sometimes called the nonlocal reduction principle was published in 1974 (see also the monograph by Gelig, Leonov and Yakubovich, 1978).

Theorem 1 *Let* $\psi : R \longrightarrow R$ *be* C^1 *and* 2π*–periodic and have exactly two zeros on an interval of the length equal to the period and such that for all* θ, $[\psi(\theta)]^2 + [\psi'(\theta)]^2 \neq 0$. *Let* $\epsilon > 0$, $\lambda > 0$, $W : R_+ \longrightarrow R$, $\sigma : R_+ \longrightarrow R$, *the functions* W, σ *being* C^1.
Assume further that

i) every solution of the equation

$$\ddot{\theta} + 2\sqrt{\lambda\epsilon}\dot{\theta} + \psi(\theta) = 0 \tag{A.0.1}$$

is bounded for $t > 0$;

ii) for all $t \in R_+$ *for which* $\psi(\sigma(t)) = 0$, $\psi'(\sigma(t)) < 0$ *we have* $W(t) \geq 0$;

iii) $$\frac{dW(t)}{dt} + 2\lambda W(t) + \epsilon\left(\frac{d\sigma(t)}{dt}^2\right) + \psi(\sigma(t))\frac{d\sigma(t)}{dt} \le 0 \qquad \text{(A.0.2)}$$

Then $\sigma(\cdot)$ *is bounded on* R_+.

Proof A. Consider the system

$$\dot{\theta} = z, \qquad \dot{z} = -2\sqrt{\lambda\epsilon}z - \psi(\theta) \qquad \text{(A.0.3)}$$

Since ψ has exactly two zeros on a period, there is one such that $\psi(\tilde{\sigma}) = 0$, $\psi'(\tilde{\sigma}) < 0$, hence $(\tilde{\sigma}, 0)$ is a saddle point for the above system; for the second zero of ψ, $\hat{\sigma} < \tilde{\sigma}$, we will have $\psi'(\hat{\sigma}) > 0$.

Since $(\tilde{\sigma}, 0)$ is a saddle point, there exists a separatrix $\tilde{\theta}_+(\cdot)$, $\tilde{\zeta}_+(\cdot)$, with $\tilde{\zeta}_+(t) > 0$ for t large and $\lim_{t\to\infty}\tilde{\theta}_+(t) = \tilde{\sigma}$, $\lim_{t\to\infty}\tilde{\zeta}_+(t) = 0$; there exists also a separatrix $\tilde{\theta}_-(\cdot)$, $\tilde{\zeta}_-(\cdot)$ with $\tilde{\zeta}_-(t) < 0$ for t large and $\lim_{t\to\infty}\tilde{\theta}_-(t) = \tilde{\sigma}$, $\lim_{t\to\infty}\tilde{\zeta}_-(t) = 0$.

We will prove that $\tilde{\zeta}_+(t) > 0$ for all t; if not, there exists t_0 such that $\tilde{\zeta}_+(t_0) = 0$, $\tilde{\zeta}_+(t) > 0$, for $t > t_0$; moreover $\tilde{\theta}_+(t_0) \in (\tilde{\sigma} - 2\pi, \tilde{\sigma})$. We know indeed that for $t > t_0$ we have $\tilde{\theta}'_+(t) = \tilde{\zeta}_+(t) > 0$ hence $t \longmapsto \tilde{\theta}_+(t)$ is increasing and $\lim_{t\to\infty}\tilde{\theta}_+(t) = \tilde{\sigma}$; therefore $\tilde{\theta}_+(t_0) < \tilde{\sigma}$.

Assume that $\tilde{\theta}_+(t_0) < \tilde{\sigma} - 2\pi$; since ψ is 2π–periodic, if $\tilde{\theta}_+(\cdot)$, $\tilde{\zeta}_+(\cdot)$ is a solution of (A.0.3) then for each $k \in Z$, $(\tilde{\theta}_+(\cdot) - 2k\pi, \tilde{\zeta}_+(\cdot))$ is also a solution and $\lim_{t\to\infty}[\tilde{\theta}_+(t) - 2k\pi, \tilde{\zeta}_+(t)] = (\tilde{\sigma} - 2k\pi, 0)$. The curve defined by $(\tilde{\theta}_+(t), \tilde{\zeta}_+(t))$, $t \in [t_0, \infty)$, together with the segment $[\tilde{\theta}_+(t_0), \tilde{\sigma}]$ of the axis $\zeta = 0$, is a simple closed curve separating the plane in two regions – the interior one and the exterior one. The solution curve defined by $(\tilde{\theta}_+(t) - 2\pi, \tilde{\zeta}_+(t))$, $t \in [t_0, \infty)$ ends at $(\tilde{\sigma} - 2\pi, 0)$ coming from the exterior (since it starts at $(\tilde{\theta}_+(t_0) - 2\pi, 0)$) and remains in the half–plane $\zeta > 0$. In this case, it must intersect $(\tilde{\theta}_+(t), \tilde{\zeta}_+(t))$, but this contradicts the uniqueness of solutions of the differential equation

$$\zeta\frac{\partial\zeta}{\partial\theta} = -2\sqrt{\lambda\epsilon}\zeta - \psi(\theta) \qquad \text{(A.0.4)}$$

in the domain $\theta \in R$, $\zeta > 0$.

In this way, we deduce that $\tilde{\theta}_+(t_0) \in (\tilde{\sigma} - 2\pi, \tilde{\sigma})$. Note now that since $\psi'(\hat{\sigma}) > 0$ we have $\psi(\theta) > 0$ for $\hat{\sigma} < \theta < \tilde{\sigma}$ and in the points $(\theta, 0)$ with $\hat{\sigma} < \theta < \tilde{\sigma}$ the vector field oriented towards the half-plane $\zeta < 0$; we deduce that $\tilde{\theta}_+(t_0) < \hat{\sigma}$ (otherwise it could not increase!). Every solution which starts at $\zeta_0 = 0$ with $\tilde{\sigma} - 2\pi < \theta_0 < \tilde{\theta}_+(t_0)$ cannot intersect $\zeta = 0$ neither between $\tilde{\sigma} - 2\pi$ and $\tilde{\theta}_+(t_0)$ because of the orientation of the vector field, nor between $\tilde{\theta}_+(t_0)$ and $\tilde{\sigma}$ because otherwise it would have to intersect the solution curve $(\tilde{\theta}_+(\cdot), \tilde{\zeta}_+(\cdot))$.

It follows that such solution has to have the component θ unbounded, and we would contradict assumption i) of the Theorem. We have thus proved that $\tilde{\zeta}_+(t) > 0$.

We prove now that $\lim_{t\to-\infty} \tilde{\zeta}_+(t) = +\infty$. If $\tilde{\zeta}_+(\cdot)$ were bounded, $\beta > 0$ would exist such that $\tilde{\zeta}_+(t) < \beta$ for $t < 0$ and the same would be true for all solutions $\left(\tilde{\theta}_+(\cdot) + 2k\pi, \tilde{\zeta}_+(\cdot)\right)$; every solution with $\zeta(0) > \beta$ would be unbounded since otherwise it would have to intersect one of the separatrices.

Let us prove now that there exists $\tilde{t} < 0$ such that

$$\tilde{\zeta}_+(\tilde{t}) = \frac{1}{2\sqrt{\lambda\epsilon}} \left[1 + \max|\psi(\sigma)|\right],$$

$$\tilde{\zeta}_+(t) < \frac{1}{2\sqrt{\lambda\epsilon}} \left[1 + \max|\psi(\sigma)|\right] \quad \text{for } t < \tilde{t}.$$

If not, there would exist $\hat{t} < \tilde{t}$ such that

$$\tilde{\zeta}_+(\hat{t}) = \frac{1}{2\sqrt{\lambda\epsilon}} \left[1 + \max|\psi(\sigma)|\right], \qquad \frac{d}{dt}\tilde{\zeta}_+(\hat{t}) \geq 0.$$

But

$$\frac{d}{dt}\tilde{\zeta}_+(\hat{t}) = -2\sqrt{\lambda\epsilon}\tilde{\zeta}_+(\hat{t}) - \psi(\tilde{\theta}_+(\hat{t})) =$$

$$= -1 - \psi(\tilde{\theta}_+(\hat{t})) - \max|\psi(\sigma)| \leq -1.$$

We have now

$$\frac{d}{dt}\tilde{\zeta}_+(t) \leq -1 \qquad \text{for } t < \tilde{t},$$

$$\int_t^{\tilde{t}} \frac{d}{d\tau}\tilde{\zeta}_+(\tau)d\tau \le -(\tilde{t}-t) \qquad \text{for } t<\tilde{t},$$

$$\tilde{\zeta}_+(\tilde{t}) - \tilde{\zeta}_+(t) \le -(\tilde{t}-t), \qquad \tilde{\zeta}_+(t) \ge \tilde{\zeta}_+(\tilde{t}) + \tilde{t} - t$$

and $\lim_{t\to-\infty} \tilde{\zeta}_+(t) = +\infty$.

In the same way, for the solution $(\tilde{\theta}_-(\cdot), \tilde{\zeta}_-(\cdot)$, we deduce that $\tilde{\zeta}_-(t) < 0$ for all t and $\lim_{t\to-\infty} \tilde{\zeta}_-(t) = -\infty$.

Since $\frac{d}{dt}\tilde{\theta}_-(t) = \tilde{\zeta}_-(t) < 0$ the function $\tilde{\theta}_-(\cdot)$ is decreasing and since $\lim_{t\to-\infty} \tilde{\zeta}_-(t) = -\infty$ we have for $t < -\hat{t}$ that $\tilde{\zeta}_-(t) < -1$, hence

$$\int_t^{-\hat{t}} \frac{d}{d\tau}\tilde{\theta}_-(\tau)d\tau < t + \hat{t}, \qquad \tilde{\theta}_-(-\hat{t}) - \tilde{\theta}_-(t) < t + \hat{t},$$

$$\tilde{\theta}_-(t) > \tilde{\theta}_-(\hat{t}) - t - \hat{t}, \qquad \lim_{t\to-\infty} \tilde{\theta}_-(t) = +\infty$$

In a similar way we deduce that $\lim_{t\to-\infty} \tilde{\theta}_+(t) = -\infty$.

B. For $k \in Z$ we will construct now a certain function $F_k(\theta)$.

We start with $(\tilde{\theta}_+(\cdot), \tilde{\zeta}_+(\cdot))$, and $(\tilde{\theta}_-(\cdot), \tilde{\zeta}_-(\cdot))$ and we associate with them $(\tilde{\theta}_+(\cdot) - 2k\pi, \tilde{\zeta}_+(\cdot))$ and $(\tilde{\theta}_-(\cdot) - 2k\pi, \tilde{\zeta}_-(\cdot))$. The functions $t \longmapsto \tilde{\theta}_+(t) - 2k\pi$, $t \longmapsto \tilde{\theta}_-(t) - 2k\pi$ are strictly monotone, hence they are invertible; let $\tilde{\tau}_+^k : (-\infty, \tilde{\sigma} - 2k\pi) \longrightarrow R$, $\tilde{\tau}_-^k : (\tilde{\sigma} - 2k\pi, \infty) \longrightarrow R$ be the inverses; we have

$$\lim_{\theta\to-\infty} \tilde{\tau}_+^k(\theta) = -\infty, \qquad \lim_{\theta\to\tilde{\sigma}-2k\pi} \tilde{\tau}_+^k(\theta) = +\infty,$$

$$\lim_{\theta\to\tilde{\sigma}-2k\pi} \tilde{\tau}_-^k(\theta) = +\infty, \qquad \lim_{\theta\to+\infty} \tilde{\tau}_-^k(\theta) = -\infty$$

Define

$$\tau_k(\theta) = \begin{cases} \tilde{\tau}_+^k(\theta), & -\infty < \theta < \tilde{\sigma} - 2k\pi \\ \tilde{\tau}_-^k(\theta), & \tilde{\sigma} - 2k\pi < \theta < +\infty \end{cases}$$

$$F_k(\theta) = \begin{cases} \tilde{\zeta}_+(\tau_k(\theta)), & -\infty < \theta < \tilde{\sigma} - 2k\pi \\ \tilde{\zeta}_-(\tau_k(\theta)), & \tilde{\sigma} - 2k\pi < \theta < +\infty \end{cases}$$

It is clear that F_k are continuous and differentiable for $\theta \in (-\infty, \tilde{\sigma} - 2k\pi)$ and $\theta \in (\tilde{\sigma} - 2k\pi, \infty)$. For $\theta \in (-\infty, \tilde{\sigma} - 2k\pi)$ we have

$$F_k'(\theta) = \left[\frac{d}{dt}\tilde{\zeta}_+(\tau_k(\theta))\right] \tau_k'(\theta) = \frac{\frac{d}{dt}\tilde{\zeta}_+(\tau_k(\theta))}{\tilde{\zeta}_+(\tau_k(\theta))} =$$

$$= -2\sqrt{\lambda\epsilon} - \frac{\psi(\tilde{\theta}_+^k(\tau_k(\theta)))}{\tilde{\zeta}_+(\tau_k(\theta))}, \qquad \tilde{\theta}_+^k(\cdot) = \tilde{\theta}_+(\cdot) - 2k\pi.$$

We deduce that

$$F_k(\theta)F_k'(\theta) + 2\sqrt{\lambda\epsilon}F_k(\theta) + \psi(\theta) = 0 \tag{A.0.5}$$

and the same formula holds for $\theta \in (\tilde{\sigma} - 2k\pi, \infty)$.

From the definition, we see that $\tau_k(\theta) = \tau_0(\theta + 2k\pi)$, hence $F_k(\theta) = F_0(\theta + 2k\pi)$. Again from the definition we see that $F_k(\tilde{\sigma} - 2k\pi) = F_k(\tilde{\sigma}) = 0$. ($\lim_{\theta \nearrow \tilde{\sigma}} F_0(\theta) = \lim_{\theta \nearrow \tilde{\sigma}} \tilde{\zeta}_+(\tau_0(\theta)) = \lim_{t \to \infty} \tilde{\zeta}_+(t) = 0$ and a similar argument holds for $\theta \searrow \tilde{\sigma}$).

Finally, from the properties of $\tilde{\zeta}_+$ and $\tilde{\zeta}_-$, we see that

$$\lim_{\theta \to \pm\infty} [F_k(\theta)]^2 = +\infty$$

C. Let W be the function in the statement and let us define

$$V_k(t) = W(t) - \frac{1}{2}(F_k(\sigma(t)))^2 \tag{A.0.6}$$

We have

$$\frac{dV_k(t)}{dt} + 2\lambda V_k(t) = \frac{dW(t)}{dt} - F_k(\sigma(t))F_k'(\sigma(t))\frac{d\sigma(t)}{dt} +$$

$$2\lambda W(t) - \lambda(F_k(\sigma(t)))^2 \leq -\epsilon\left(\frac{d\sigma(t)}{dt}\right)^2 + 2\sqrt{\lambda\epsilon}F_k(\sigma(t))\frac{d\sigma(t)}{dt} -$$

$$-\lambda(F_k(\sigma(t)))^2 = -\left[\sqrt{\epsilon}\frac{d\sigma(t)}{dt} - \sqrt{\lambda}F_k(\sigma(t))\right]^2$$

Let $k > 0$ be such that $|\sigma(0) - \tilde{\sigma}| \leq 2k\pi$; we may choose k large enough to have also

$$V_k(0) = W(0) - \frac{1}{2}(F_k(\sigma(0)))^2 =$$

$$= W(0) - \frac{1}{2}[F_0(\sigma(0) + 2k\pi)]^2 < 0$$

$$V_{-k}(0) = W(0) - \frac{1}{2}[F_0(\sigma(0) - 2k\pi)]^2 < 0.$$

We have

$$\frac{dV_k(t)}{dt} + 2\lambda V_k(t) \leq 0$$

hence $e^{2\lambda t}V_k(t) \leq V_k(0)$, and therefore $V_k(t) < 0$, $V_{-k}(t) < 0$ for all $t \geq 0$.

We deduce from here that for all $t \geq 0$ we have $|\sigma(t) - \tilde{\sigma}| \leq 2k\pi$. If not, there were $\hat{t} > 0$ such that $\sigma(\hat{t}) - \tilde{\sigma} = 2k\pi$. Since ψ is 2π–periodic $\psi(\sigma(\hat{t})) = \psi(\tilde{\sigma} + 2k\pi) = 0$, $\psi'(\sigma(\hat{t})) = \psi'(\tilde{\sigma}) < 0$ and $W(\hat{t}) \geq 0$. We will also have either $F_k(\sigma(\hat{t})) = 0$ or $F_{-k}(\sigma(\hat{t})) = 0$ and we deduce that either $V_k(\hat{t}) \geq 0$ or $V_{-k}(\hat{t}) \geq 0$ which isa contradiction.

The inequality $|\sigma(t) - \tilde{\sigma}| \leq 2k\pi$ shows that $\sigma(\cdot)$ is bounded and this completes the proof.

Appendix 2
A Second Order Equation

We will consider the equation

$$\ddot{\theta} + k\dot{\theta} + \sin\theta + r\sin 2\theta = P \tag{A.0.7}$$

which is obtained from the simplest model for a synchronous machine if

$$\delta(t) = \theta\left(t\sqrt{\frac{e_f U\omega}{x_d T}}\right), \qquad r = \left(\frac{1}{x_q} - \frac{1}{x_d}\right)\frac{x_d U}{2e_f},$$

$$P = \frac{P_{mec} x_d}{e_f U}, \qquad k = D\sqrt{\frac{x_d}{e_f U \omega T}}.$$

With (A.0.7) we associate the system

$$\dot{\theta} = z = \Theta(\theta, z)$$

$$\dot{z} = P - \sin\theta - r\sin 2\theta - kz = Z(\theta, z). \tag{A.0.8}$$

Let us note that a natural phase space for this dynamical system is a cylinder since θ is an angular variable.

The equilibria are obtained for

$$z = 0, \qquad \sin\theta + r\sin 2\theta = P.$$

Defining $f : R \longrightarrow R$ by $f(\theta) = \sin\theta + r\sin 2\theta$, we have

$$f'(\theta) = \cos\theta + 2r\cos 2\theta = \cos\theta + 2r(2\cos^2\theta - 1) =$$

$$= 4r\cos^2\theta + \cos\theta - 2r = 4r(\cos^2\theta + \frac{1}{4r}\cos\theta - \frac{1}{2}) =$$

$$= 4r\left[\left(\cos\theta + \frac{1}{8r}\right)^2 - \left(\frac{1}{64r^2} + \frac{1}{2}\right)\right].$$

Let

$$\xi_2(r) = -\frac{1}{8r} + \frac{\sqrt{1+32r^2}}{8r} > 0.$$

Note that $\xi_2(r) < 1$ hence there exist $\theta_1(r)$, $\theta_2(r)$ such that

$$f'(\theta_1(r)) = f'(\theta_2(r)) = 0, \quad \theta_1(r) \in [0, 2\pi), \quad \theta_2(r) \in [0, 2\pi)$$

If we consider also

$$\xi_1(r) = -\frac{1}{8r} - \frac{\sqrt{1+32r^2}}{8r}$$

we see that $\xi_1(r) \geq -1$ if and only if $r \geq \frac{1}{2}$; in this case we have also $\theta_3(r)$, $\theta_4(r)$ located in $[0, 2\pi)$ such that $f'(\theta_3(r)) = f'(\theta_4(r)) = 0$.

Notice that $\theta_2(\tau) = 2\pi - \theta_1(\tau)$, $0 \leq \theta_1(\tau) \leq \frac{\pi}{2}$, $\theta_1(\tau)$ is a maximum for f while $\theta_2(\tau)$ is a minimum.

Denote $P_1(\tau) = f(\theta_1(\tau))$; if $P < P_1(\tau)$ the equation $f(\theta) = P$ will have two solutions located in $(0, \pi)$ corresponding to two equilibria.

If $\tau \geq \frac{1}{2}$ we will have also the zeros of f' at $\theta_3(\tau)$ and $\theta_4(\tau) = 2\pi - \theta_3(\tau)$ If $P > f(\theta_4(\tau))$ we have again only two equilibria, while if $P > f(\theta_4(\tau))$ there will be four equilibria. If $P > f(\theta_1(\tau))$ there are no equilibria at all. Let us note that usually the situation with four equilibria is avoided: $\tau \geq \frac{1}{2}$ corresponds to small e_f that is an underexcited machine while $P < f(\theta_4(\tau))$ means a low load; the machine is not usually exploited at low load and small excitation.

We will be interested in the situation where there are exactly two equilibria in $[0, 2\pi)$ since this is one of the assumptions we had in the reduction principle (Appendix 1).

Let us note that for $f'(\hat{\theta}) > 0$ the equilibrium is stable, while for $f'(\hat{\theta}) < 0$ it is a saddle point.

Consider now the function g defined by

$$g(\theta) = \frac{1}{k}[P - f(\theta)].$$

From the above analysis it follows that under our assumptions g has on a period a maximum and a minimum; the maximal value is positive and is denoted by z_M. If we consider the line $z = C_1 > z_M$; along this line we have $\frac{dz}{dt} < 0$, $\frac{d\theta}{dt} > 0$ and we deduce that for all $t > 0$ the z component must be bounded from above. On the other hand, on the line $z = C_2 < \min(0, z_m)$, where z_m is a minimal value of g, we will have $\frac{dz}{dt} > 0$, $\frac{d\theta}{dt} < 0$ and as a result z is also bounded from below. In this way the boundedness assumption is always fulfilled for z and we have to look closer at the behaviour in θ only. We will consider only the situation with two equilibria $\tilde{\theta}_1(\tau, P)$, $\tilde{\theta}_2(\tau, P)$ with $0 < \tilde{\theta}_1(\tau, P) < \tilde{\theta}_2(\tau, P) < 2\pi$, $f'(\tilde{\theta}_1(\tau, P)) > 0$, $f'(\tilde{\theta}_2(\tau, P)) < 0$. We will have boundedness of θ if every trajectory ends at an equilibrium. To have such a behaviour, the separatrix entering $(\tilde{\theta}_1(\tau,P),0)$ located in the half - plane $z>0$, must cut the line $\theta=\tilde{\theta}_1(\tau,P)$ above the point where this line intersects the separatrix starting at $(\tilde{\theta}_2(\tau,P)-2\pi,0)$. We will consider the curve

$$z(\theta)=\sqrt{F(\theta)-F(\tilde{\theta}_2(\tau,P))+k^2(\tilde{\theta}_2(\tau,P)-\theta)^2}$$

with $F(\theta)=2P\theta+2\cos\theta+2\cos 2\theta$. We have

$$z'(\theta)=\frac{F'(\theta)-2k^2(\tilde{\theta}_2(\tau,P)-\theta)}{\sqrt{F(\theta)-F(\tilde{\theta}_2(\tau,P))+k^2(\tilde{\theta}_2(\tau,P)-\theta)^2}}$$

$$F'(\theta)=2(P-f(\theta)),\qquad F''(\theta)=-2f'(\theta),$$

$$F'(\tilde{\theta}_2(\tau,P))=0$$

$$F'(\theta)=F''(\tilde{\theta}_2(\tau,P))(\theta-\tilde{\theta}_2(\tau,P))+o(\theta-\tilde{\theta}_2(\tau,P))$$

$$F(\theta)-F(\tilde{\theta}_2(\tau,P))+k^2(\tilde{\theta}_2(\tau,P)-\theta)^2=$$
$$\left[\frac{1}{2}F''(\tilde{\theta}_2(\tau,P))+k^2\right](\tilde{\theta}_2(\tau,P)-\theta)^2+O((\tilde{\theta}_2(\tau,P)-\theta)^3).$$

Hence

$$z'(\theta)=\frac{2[f'(\tilde{\theta}_2(\tau,P))-k^2](\tilde{\theta}_2(\tau,P)-\theta)+o(\tilde{\theta}_2(\tau,P)-\theta)}{2\sqrt{k^2-f'(\tilde{\theta}_2(\tau,P))+O(\tilde{\theta}_2(\tau,P)-\theta)}(\tilde{\theta}_2(\tau,P)-\theta)}.$$

Therefore

$$\lim_{\theta\to\tilde{\theta}_2(\tau,P)} z'(\theta)=-\sqrt{k^2-f'(\tilde{\theta}_2(\tau,P))}.$$

On the other hand, the tangent to the separatix entering $(\tilde{\theta}_2(\tau,P),0)$ at this point is

$$z=-\left(\frac{k}{2}+\sqrt{\frac{k^2}{4}-f'(\tilde{\theta}_2(\tau,P))}\right)(\theta-\tilde{\theta}_2(\tau,P)).$$

We have

$$\frac{k}{2}+\sqrt{\frac{k^2}{4}-f'(\tilde{\theta}_2(\tau,P))}>\sqrt{k^2-f'(\tilde{\theta}_2(\tau,P))}$$

hence the slope on the separatrix is smaller than the one on the comparison curve; we deduce that the comparison curve is located below the separatrix.

The comparison curve is contact–free in the interval $\left(\tilde{\theta}_1(\tau,P),\tilde{\theta}_2(\tau,P)\right)$. Indeed, in this interval we have $P-f(\theta)<0$, hence $F'(\theta)<0$, hence $F(\theta)>F(\tilde{\theta}_2(\tau,P))$ and $z(\theta)>k[\tilde{\theta}_2(\tau,P)-\theta]$, $z(\theta)z'(\theta)=P-f(\theta)-k^2[\tilde{\theta}_2(\tau,P)-\theta]$, $z(\theta)z'(\theta)<P-f(\theta)-kz(\theta)$, while on integral curves we have equality.

We deduce that the point $(\tilde{\theta}_1(\tau,P),h_1)$ on the separatrix entering $(\tilde{\theta}_2(\tau,P),0)$ is such that

$$h_1>\sqrt{F(\tilde{\theta}_1(\tau,P))-F(\tilde{\theta}_2(\tau,P))+k^2(\tilde{\theta}_1(\tau,P)-\tilde{\theta}_2(\tau,P))^2}$$

For the separatrix starting at $(\tilde{\theta}_2(\tau,P)-2\pi,0)$ in the half plane $z>0$, we consider also, as a comparison curve, the trajectory corresponding to $k=0$, defined by

$$z(\theta)=\sqrt{F(\theta)-F(\tilde{\theta}_2(\tau,P)-2\pi)}.$$

This is also contact–free and is located above the separatrix; we deduce that the point $(\tilde{\theta}_1(\tau,P),h_2)$ on this separatrix is such that

$$h_2<\sqrt{F(\tilde{\theta}_1(\tau,P))-F(\tilde{\theta}_2(\tau,P)-2\pi)}$$

If $h_2<h_1$ a simple uniqueness argument shows that all trajectories have to end in a singular point. We deduce that a sufficient condition for all solutions to our equation to be bounded is that

$$F(\tilde{\theta}_1(\tau,P))-F(\tilde{\theta}_2(\tau,P)-2\pi)\le F(\tilde{\theta}_1(\tau,P))-F(\tilde{\theta}_2(\tau,P))+$$
$$+k^2\left[\tilde{\theta}_1(\tau,P)-\tilde{\theta}_2(\tau,P)\right]^2.$$

But

$$F(\tilde{\theta}_2(\tau,P)-2\pi)=2P[\tilde{\theta}_2(\tau,P)-2\pi]+2\cos\tilde{\theta}_2(\tau,P)+$$

$$+r\cos\tilde{\theta}_2(r,P)$$

and our inequality reads

$$k^2\left[\tilde{\theta}_2(r,P)-\tilde{\theta}_1(r,P)\right]^2 \geq 4P\pi$$

that is

$$k \geq \frac{2\sqrt{P\pi}}{\tilde{\theta}_2(r,P)-\tilde{\theta}_1(r,P)}. \tag{A.0.9}$$

It is worth mentioning that this sufficient condition can be improved by a better choice of the comparison curves. At this point a long list of references can be mentioned (e.g. Amerio, 1949; Seifert, 1952, 1953, 1959; Bohm, 1953; Hayes, 1953; Barbălat and Halanay, 1959; Barbašin and Tabueva, 1969).

Appendix 3
Liapunov Equations

Proposition 1 *If the matrices* A, B *have no common eigenvalues, then for all matrices* C *the equation*

$$AX - XB = C \tag{A.0.10}$$

has a unique solution.

Proof We have to prove that the linear equation $AX - XB = 0$ has only the solution $X = 0$. For every solution of this equation we have

$$(-\sigma I + A)X + X(\sigma I - B) = 0,$$

$$X(\sigma I - B)^{-1} = (\sigma I - A)^{-1}X.$$

The functions $\sigma \longmapsto X(\sigma I - B)^{-1}$ and $\sigma \longmapsto (\sigma I - A)^{-1}X$ are rational and since A and B have no common eigenvalues, the function $\sigma \longmapsto X(\sigma I - B)^{-1} = (\sigma I - A)^{-1}X$ has no poles hence by the Liouville theorem must be constant. On the other hand it tends to zero at the infinity hence $X(\sigma I - B)^{-1} = (\sigma I - A)^{-1}X = 0$ and we deduce $X = 0$.

Proposition 2 *If A is a Hurwitz matrix then the equation*

$$A^*X + XA = -Q \tag{A.0.11}$$

has a unique solution given by

$$X = \int_0^\infty e^{A^*t} Q e^{At} dt. \tag{A.0.12}$$

Thus, if $Q \geq 0$, *then* $X \geq 0$.

Proof We have only to check the formula for the solution

$$A^*X + XA = \int_0^\infty (A^* e^{A^*t} Q e^{At} + e^{A^*t} Q e^{At} A) dt =$$

$$= \int_0^\infty \frac{d}{dt} \left(e^{A^*t} Q e^{At} \right) dt = -Q,$$

since $\lim_{t \to \infty} e^{At} = 0$ because A is a Hurwitz matrix.

Corollary 1 *Let* P *be symmetric and* A *be such that* $A^*P + PA < 0$; *then the necessary and sufficient condition for* A *to be Hurwitz is that* $P > 0$.

Proof If A is a Hurwitz matrix and if we denote $-Q = A^*P + PA$, we have by assumption $Q > 0$, hence from the above formula $P \geq 0$. Now

$$x^*Px = \int_0^\infty x^* e^{A^*t} Q e^{At} x dt$$

and if $x^*Px = 0$ then $x^* e^{A^*t} Q e^{At} x \equiv 0$ hence $e^{At}x \equiv 0$ that is $x = 0$ and we see that $P > 0$.

Let now $P > 0$, λ be an eigenvalue for A, u a corresponding eigenvector; we have $Au = \lambda u$, $u \neq 0$;

$$u^*(A^*P + PA)u = u^*\bar{\lambda}Pu + u^*P\lambda u = (\lambda + \bar{\lambda})u^*Pu < 0$$

and since $u^*Pu > 0$ it follows that $\lambda + \bar{\lambda} < 0$ hence $\Re\lambda < 0$.

Proposition 3 *Let* A *be a matrix whose eigenvalues are purely imaginary and having simple elementary divisors. Then there exists a positive definite symmetric matrix* P *such that*

$$A^*P + PA = 0$$

Proof Under our assumptions, there exists an invertible matrix S such that $SA^{-1}S$ is diagonal with purely imaginary entries. Denoting $SA^{-1}S = D$, we see that $D + D^* = 0$. Let $\tilde{X}$ be an arbitrary diagonal matrix with strictly positive entries; we have $\tilde{X}D + D^*\tilde{X} = \tilde{X}(D + D^*) = 0$. Take $P = S^*\tilde{X}S$; since $\tilde{X} > 0$ and S is invertible we have $P > 0$. Further

$$A^*P + PA = A^*S^*\tilde{X}S + S^*\tilde{X}SA =$$

$$= S^*\tilde{X}DS + S^*D^*\tilde{X}S = S^*(\tilde{X}D + D^*\tilde{X})S = 0.$$

Appendix 4
The Yakubovich – Kalman – Popov Lemma

Here, we will present a simple proof of the lemma for the specific case in which we have used it.

Lemma 1 *Let* A *be a Hurwitz matrix,* (A, b) *controllable and let*

$$\chi(\lambda, \sigma) = \kappa + c^*(\sigma I - A)^{-1}b + b^*(\lambda I - A^*)^{-1}c +$$

$$+ b^*(\lambda I - A^*)^{-1}M(\sigma I - A)^{-1}b$$

If

$$\chi(-\imath\omega, \imath\omega) \geq 0\,, \quad \omega \in R \tag{A.0.13}$$

then there exist γ, w, N, $N = N^*$ *such that*

$$\gamma\bar{\gamma} = \kappa\,, \quad c + Nb = \gamma w\,, \quad M + NA + A^*N = ww^* \tag{A.0.14}$$

Proof **A**. We will prove that there exist γ and w such that

$$\chi(-\sigma,\sigma) = \overline{\varphi(-\bar{\sigma})}\varphi(\sigma), \qquad \varphi(\sigma) = \gamma + w^*(\sigma I - A)^{-1}b$$

We define

$$\Pi(-\sigma,\sigma) = \frac{1}{\nu}\det(-\sigma I - A^*)\det(\sigma I - A)\chi(-\sigma,\sigma)$$

where $\nu = \max|\alpha_j|$, α_j being the coefficients of the polynomial

$$\det(-\sigma I - A^*)\det(\sigma I - A)\chi(-\sigma,\sigma).$$

We have

$$\Pi(-\imath\omega,\imath\omega) = \frac{1}{\nu}|\det(\imath\omega I - A)|^2\chi(-\imath\omega,\imath\omega) \geq 0.$$

If Π is constant, this constant is positive and we can write

$$\Pi(-\sigma,\sigma) = \bar{\alpha}_0\alpha_0\,,$$

$$\chi(-\sigma,\sigma) = \frac{\bar{\alpha}_0}{\det(-\sigma I - A^*)}\cdot\frac{\alpha_0}{\det(\sigma I - A)}.$$

Let us note now now that for all w we have $w^*(\sigma I - A)^{-1}b = \frac{\Pi(\sigma)}{\det(\sigma I - A)}$, where $\Pi(\sigma)$ is a polynomial of degree at most $n-1$. If we assume (A,b) to be controllable, w is uniquely determined by $\Pi(\sigma)$; it follows that there exists w such that

$$w^*(\sigma I - A)^{-1}b = \frac{\alpha_0}{\det(\sigma I - A)}.$$

Using such w, we have

$$\chi(-\sigma,\sigma) = \overline{\varphi(-\bar{\sigma})}\varphi(\sigma), \qquad \varphi(\sigma) = w^*(\sigma I - A)^{-1}b$$

This simple case shows how one should proceed in the general one. Suppose we obtained $\Pi(-\sigma,\sigma) = \overline{\psi(-\bar{\sigma})}\psi(\sigma)$; take

$$\varphi(\sigma) = \sqrt{\nu}\frac{\psi(\sigma)}{\det(\sigma I - A)}$$

then $\overline{\varphi(-\bar{\sigma})}\varphi(\sigma) = \chi(-\sigma, \sigma)$.
On the other hand, $\sqrt{\nu}\psi(\sigma)$ is a polynomial of degree at most n; define

$$\gamma = \lim_{\sigma\to\infty} \sqrt{\nu}\frac{\psi(\sigma)}{\det(\sigma I - A)}$$

Then $\tilde{\psi}(\sigma) = \sqrt{\nu}\psi(\sigma) - \gamma \det(\sigma I - A)$ is of degree less than n and from the controllability we deduce that there exists w such that

$$w^*(\sigma I - A)^{-1}b = \frac{\tilde{\psi}(\sigma)}{\det(\sigma I - A)} = \varphi(\sigma) - \gamma$$

We have still to derive the factorization formula

$$\Pi(-\sigma, \sigma) = \overline{\psi(-\bar{\sigma})}\psi(\sigma)$$

We can assume that Π is not constant since we have already considered that case. Let σ_1 be such that $\Pi(-\sigma_1, \sigma_1) = 0$. Notice now that

$$\overline{\Pi(\bar{\sigma}, -\bar{\sigma})} = \frac{1}{\nu}\overline{\det(\bar{\sigma} I - A^*)}\,\overline{\det(-\sigma I - A)}\,\overline{\chi(\bar{\sigma}, -\bar{\sigma})} =$$

$$= \frac{1}{\nu}\det(\sigma I - A)\det(-\bar{\sigma} I - A^*)\left[\kappa + \overline{c^*(-\bar{\sigma} I - A)^{-1}b} + \right.$$

$$\left. + \overline{b^*(\bar{\sigma} I - A^*)^{-1}c} + \overline{b^*(\bar{\sigma} I - A^*)^{-1}M(-\bar{\sigma} I - A)^{-1}b}\right] =$$

$$= \Pi(-\sigma, \sigma).$$

We deduce that $\overline{\Pi(-\bar{\sigma}_1, \bar{\sigma}_1)} = 0$ hence $\Pi(-\bar{\sigma}_1, \bar{\sigma}_1) = 0$.If $\sigma_1 = -\bar{\sigma}_1$, σ_1 is purely imaginary and since $\Pi(-\imath\omega, \imath\omega) \geq 0$, σ_1 must have even multiplicity. We find that we can write

$$\Pi(-\sigma, \sigma) = (\sigma - \sigma_1)(\sigma + \bar{\sigma}_1)Q(\sigma)$$

$$Q(\sigma) = \frac{\Pi(-\sigma, \sigma)}{(\sigma - \sigma_1)(\sigma + \bar{\sigma}_1)}$$

$$\overline{Q(-\bar{\sigma})} = \frac{\overline{\Pi(\bar{\sigma}, -\bar{\sigma})}}{\overline{(-\bar{\sigma} - \sigma_1)(-\bar{\sigma} + \bar{\sigma}_1)}} =$$

$$= \frac{\Pi(-\sigma,\sigma)}{-(\sigma+\bar{\sigma}_1)(-\sigma+\sigma_1)} = \frac{\Pi(-\sigma,\sigma)}{(\sigma-\sigma_1)(\sigma+\bar{\sigma}_1)}$$

and Q has the same property as $\sigma \longmapsto \Pi(-\sigma,\sigma)$ but its degree is less by 2. We have further

$$Q(\imath\omega) = \frac{\Pi(-\imath\omega,\imath\omega)}{(\imath\omega+\bar{\sigma}_1)(-\imath\omega-\sigma_1)} = -\frac{\Pi(-\imath\omega,\imath\omega)}{|\imath\omega+\sigma_1|^2}$$

and we can proceed with $-Q(\sigma)$ precisely as we did for $\Pi(-\sigma,\sigma)$.

As a final result, we obtain

$$\Pi(-\sigma,\sigma) = \gamma_0\bar{\gamma}_0(\sigma-\sigma_1)(\sigma-\sigma_2)\ldots(\sigma-\sigma_p)(\sigma+\bar{\sigma}_1)\ldots(\sigma+\bar{\sigma}_p)$$

and we may choose

$$\psi(\sigma) = \gamma_0(\sigma-\sigma_1)\ldots(\sigma-\sigma_p)$$

Let us note that we have freedom to chose appropriate ψ in order to get a Hurwitz polynomial.

B. It is clear from what was said above that $\gamma\bar{\gamma} = \kappa$. On the other hand, we assumed that A is a Hurwitz matrix. From the Liapunov theorem it follows that there exists N such that

$$NA + A^*N = ww^* - M.$$

To complete the proof, we have to prove that $c + Nb = \gamma w$. The first computation leads to

$$b^*(-\sigma I - A^*)^{-1}ww(\sigma I - A)^{-1}b - b^*(-\sigma I - A^*)^{-1}M\cdot$$

$$\cdot(\sigma I - A)^{-1}b = -b^*(-\sigma I - A^*)^{-1}Nb - b^*N(\sigma I - A)^{-1}b.$$

On the other hand

$$\chi(-\sigma,\sigma) = \bar{\gamma}\gamma + \bar{\gamma}w^*(\sigma I - A)^{-1}b + \gamma b^*(-\sigma I - A^*)^{-1}w +$$

$$+ b^*(-\sigma I - A^*)^{-1}ww^*(\sigma I - A)^{-1}b$$

and we deduce that

$$\Re(c^* - \bar{\gamma}w^* + b^*N)(\imath\omega I - A)^{-1}b = 0$$

Consider the function $\psi_0(z) = q^*(zI - A)^{-1}b$, $q \neq 0$. This function is rational and since (A, b) is controllable it is not identically zero; since A is Hurwitz, the poles are located in the region $\Re z < 0$. Since the degree of the denominator is not zero there exist at least one pole.

let us now assume that $\Re\psi_0(z) \equiv 0$. The function $\psi(z) = \imath\psi_0(\imath z)$ is real for real z since $\psi_0(\imath z)$ is purely imaginary. We deduce that all coefficients can chosen be *real.* Now, if α is a pole, then $\bar{\alpha}$ is also a pole, but then $\imath\alpha$ and $\imath\bar{\alpha}$ are poles for ψ_0 and one of them has to be in $\Re z \geq 0$, a contradiction.

We deduce that $c^* + b^*N - \bar{\gamma}w^* = 0$ and the proof of the lemma is completed.

Appendix 5
A Result Concerning Exponential Stability

Here, we will discuss a result of A.Halanay (1960, 1966) which we employed in the proof that the linear feedback control we used to stabilize the steam turbine, leads to exponential stability.

Proposition 1 *Consider the system*

$$x' = f(x), \qquad f(0) = 0 \tag{A.0.15}$$

The zero solution is asymptotically stable if and only if there exist a continuous increasing function φ with $\varphi(0) = 0$ and a decreasing function σ tending to zero at infinity such that for all solutions with $|x_0| < \delta_0$,

$$|x(t, x_0)| \leq \varphi(|x_0|)\sigma(t), \quad t > 0 \tag{A.0.16}$$

Proof It is obvious that (A.0.16) implies asymptotic stability. On the other hand, asymptotic stability implies existence of $\delta(\epsilon)$ and $T(\epsilon)$ such that $|x_0| < \delta(\epsilon)$ implies $|x(t, x_0)| < \epsilon$ for all $t > 0$ and $|x_0| < \delta_0$

implies $|x(t,x_0)| < \epsilon$ for $t > T(\epsilon)$. Consider a sequence $(\epsilon_k)_{k\geq 0}$, $\epsilon_k > 0$, $\epsilon_k > \epsilon_{k+1}$, $\lim_{k\to\infty} \epsilon_k = 0$.

Define

$$\delta_k = \sup\{\delta \,:\, |x_0| < \delta \text{ implies } |x(t,x_0)| < \epsilon_k,\ \forall t > 0\}$$

It is clear that $\epsilon_{k+1} < \epsilon_k$ implies $\delta_{k+1} < \delta_k$. Define $\tilde{\delta}(\epsilon)$ to be linear between ϵ_{k+1} and ϵ_k and $\tilde{\delta}(\epsilon_k) = \delta_{k+1}$, where we have kept from the sequence $(\delta_k)_{k\geq 0}$ only the distinct terms, without changing notation. We obtain the function $\tilde{\delta}(\cdot)$ continuous and strictly increasing with $\tilde{\delta}(0) = 0$. If $|x_0| < \tilde{\delta}(\epsilon)$ we have $|x_0| < \tilde{\delta}(\epsilon_k) = \delta_{k+1}$ for $\epsilon_{k+1} \leq \epsilon \leq \epsilon_k$ hence $|x(t,x_0)| < \epsilon_{k+1} \leq \epsilon$; we deduce that

$$|x(t,x_0)| < \tilde{\delta}^{-1}(|x_0|),\quad t > 0$$

Further, if

$$\tau_k = \inf\{T \,:\, t > T \text{ implies } |x(t,x_0)| < \epsilon_k\}\ (|x_0| < \delta_0);$$

we have $\tau_{k+1} > \tau_k$ and define $\tilde{T}(\epsilon)$ to be linear between ϵ_k and ϵ_{k+1} and such that $\tilde{T}(\epsilon_k) = \tau_{k+1}$. We deduce that $\tilde{T}(\cdot)$ is continuous, decreasing and $\lim_{\epsilon\to 0} \tilde{T}(\epsilon) = \infty$.

It is easy to see now that for $|x_0| < \delta_0$ we can write

$$|x(t,x_0)| \leq \tilde{\delta}^{-1}(|x_0|)\tilde{T}^{-1}(t)$$

and the proof is completed.

Proposition 2 *If* $|f(x)| \leq L(\rho)|x|$ *for* $|x| < \rho$ *and*

$$|x(t,x_0)| \leq \varphi(|x_0|)\sigma(t)$$

where φ *is linear, then the zero solution of (A.0.15) is exponentially stable.*

Proof Choose T large enough to have

$$k^2\sigma^2(T) < \frac{1}{2} \text{ where } \varphi(r) = kr$$

Define

$$V(\xi) = \int_0^T |x(\tau, \xi)|^2 d\tau \tag{A.0.17}$$

Since

$$|x(\tau, \xi)| \leq k|\xi|\sigma(\tau) < k|\xi|\sigma(0)$$

we have

$$V(\xi) \leq Tk^2\sigma^2(0)|\xi|^2$$

For $|\xi| < \delta_0$ we have $|f(x(\tau, \xi))| \leq L(k\delta_0\sigma(0))|\xi|$ and from here

$$|x(\tau, \xi)|^2 \geq |\xi|^2 \exp[-2L(k\delta_0\sigma(0))\tau]$$

We deduce finally that

$$\alpha(T)|\xi|^2 \leq V(\xi) \leq \beta(T)|\xi|^2 \tag{A.0.18}$$

Define further

$$\tilde{V}(t) = V(x(t, x_0)) = \int_0^T |x(\tau, x(t, x_0))|^2 d\tau =$$

$$= \int_0^T |x(\tau + t, x_0)|^2 d\tau = \int_t^{t+T} |x(\sigma, x_0)|^2 d\sigma$$

We have

$$\tilde{V}'(t) = -|x(t, x_0)|^2 + |x(t + T, x_0)|^2$$

and, on the other hand

$$|x(T, x(t, x_0))|^2 \leq k^2|x(t, x_0)|^2\sigma^2(T) < \frac{1}{2}|x(t, x_0)|^2$$

Therefore

$$\tilde{V}'(t) \leq -\frac{1}{2}|x(t, x_0)|^2 \leq -\frac{1}{2\beta(T)}\tilde{V}(t)$$

and such inequality implies directly exponential stability

$$\tilde{V}(t) \leq e^{-\frac{t}{2\beta(T)}}\tilde{V}(0)$$

hence

$$|x(t, x_0)| \leq \sqrt{\frac{\beta(T)}{\alpha(T)}} e^{-\frac{t}{\beta(T)}} |x_0| .$$

References

ADKINS, B. (1962) *The General Theory of Electric Machines.* Chapman & Hall.

AMERIO, L. (1949) *Determinazione delle condizioni di stabilità per gli integrali di un'equazione interessante l'elettrotecnica.* Ann.Mat.Pura Appl. **4**, *30*, 475–490.

ARIE, E., BOTGROS, M., HALANAY, A., MARTAC, D. (1974) *Transient Stability of the Synchronous Machine.* Rev.Roumaine Sci.Tech.Sér.Electrotech.Energét. **19**, *4*, 611–625.

BARBALAT, I., HALANAY, A. (1959) *Evaluation de la valeur critique de l'équation généralisée du pendule.* Bull.Math.Soc. Sci. Math. Roumanie (N.S) **3(51)**, *3*, 259–275.

BARBAŠIN, E.A., TABUEVA, V.A. (1969) *Dynamic systems with cylindric phase space. Nauka* (in Russian).

BÖHM, C. (1953) *Nuovi criteri di esistenza di soluzioni periodiche di una nota equazione differenziale non lineare.* Ann. Mat.Pura Appl. **35**, *4*.

CONCORDIA, C., DE MELLO, F. (1969) *Concepts of Synchronous Machine Stability as Affected by Excitation Control.* IEEE Trans.Power Appar.Systems **PAS 88**, *4*, 316–339.

CRARY, S.B. (1947) *Power System Stability.* J. Wiley.

DRĂGAN, V., HALANAY, A. (1983) *Singular perturbations. Asymptotic expansions.* Editura Academiei (in Roumanian).

FAGIUOLI, E., SZEGÖ, G.P. (1970) *Qualitative Analysis by Modern Methods of a Stability Problem in Power System Analysis.* J.Franklin Inst. **290**, *2*.

GELIG, A.KH., LEONOV, G.A., YAKUBOVICH, V.A. (1978) *Stability of nonlinear systems with nonunique equilibrium.* Nauka (in Russian).

GOREV, A.A. (1950) *Transient processes of the synchronous machine.* Gosenergoizdat (in Russian).

HALANAY, A. (1960) *Generalization of a theorem of Persidskii.* Com.Acad.R.P.R. **12**, 1065–1068 (in Roumanian).

HALANAY, A. (1966) *Differential Equations. Stability. Oscillations. Time Lag.* Academic Press.

HALANAY, A. (1977) *Stability Problems for Synchronous Machines.* VII Int.Konf.über nichtlin.Schwing. Bd. II.1, Abh. Akad.Wiss. DDR, Abt.Math.-Natur-Tech. **5N**, 407–421.

HALANAY, A. (1978) *Stabilitá.* Pitagora Editrice Bologna.

HALANAY, A., LEONOV, G.A., RĂSVAN, VL. (1987) *From pendulum equation to an extended analysis of synchronous machines.* Rend.Sem.Mat.Univers.Politecn. Torino **45**, *2*, 91–106.

HALANAY, A., RĂSVAN, VL. (1980) *Stabilization of a Class of Bilinear Control Systems with Applications to Steam Turbine Regulation.* Tôhoku Math.Journ. **32**, *2*, 299–308.

HAYES, W.D. (1953) *On the Equation for a Damped Pendulum under Constant Torque.* Z.Angew.Math.Phys. **4**, *4*.

IEEE COMMITTEE REPORT (1973) *Dynamic Models for Steam and Hydro Turbines in Power System Studies.* IEEE Trans.Power Appar.Systems **PAS–92**, *6*, 1906–1915.

IVANOV, V.A. (1971) *Regimes of high power steam turbine units.* Energija (in Russian).

IVANOV, V.A. (1982) *Control of power units.* Mašinostroenie (in Russian).

KIRILLOV, I.I. (1988) *Automatic control of steam and gas turbines.* Mašinostroenie (in Russian).

LEONOV, G.A. (1974) *About stability of phase systems.* Sib.Mat. Ž. **15**, *1*, 49–60 (in Russian).

POPOV, V.M. (1973) *Hyperstability of Control Systems.* Springer Verlag.

QUINN, J.P. (1980) *Stabilization of Bilinear Systems by Quadratic Feedback. Controls.* J.Math.Anal.Appl. **75**, *1*, 66–80.

RĂSVAN, VL. (1981) *Stability of bilinear control systems occurring in combined heat–electricity–generation I: The mathematical models and their properties.* Rev.Roumaine Sci.Tech.Sér.Electrotech.Energét. **26**, *3*, 455–465.

RĂSVAN, VL. (1984) *Stability of bilinear control systems occurring in combined heat–electricity generation II: Stabilization of the reduced models.* Rev.Roumaine Sci.Tech.Sér.Electrotech.Energét. **29**, *4*, 423–432.

RĂSVAN, VL. (1984) *Stability of some systems with periodic nonlinearities (synchronous machines).* Rev.Roumaine Math. Pures Appl. **29**, *10*, 895–898.

SEIFERT, G. (1952) *On the Existence of Certain Solutions of Nonlinear Differential Equations.* Z.Angew.Math.Phys. **3**, *6*, 408–471.

SEIFERT, G. (1953) *On Certain Solutions of Pendulum Type Equation.* Quart.Appl.Math. **11**.

SEIFERT, G. (1959) *The Asymptotic Behavior of Solutions of Pendulum Type Equations.* Ann.Math. **69**, *1*.

SLEMROD, M. (1978) *Stabilization of Bilinear Control Systems with Application to Nonconservative Problems in Elasticity.* SIAM J.Contr.Optim. **16**, *1*, 131–141.

TOLLE, M. (1906) *Regelung der Kraftmaschinen.* Springer Verlag.

VOZNESENSKII, J.N. (1938) *About the control of the machines with large number of the control parameters* Avtom. i Telemekh. **1**, *4*, (in Russian).

CHAPTER 4

Stability Problems in Chemical Engineering

As it has been shown in the introductory part, the stability studies for chemical processes are connected with prolonged operation of chemical plants around the steady state. Chemical systems are subject to disturbances like: variations of initial concentrations of the reactor input substances, modifications of environment thermodynamic parameters, substance flow variations, control system signals.

If short–period disturbances are considered, their effect can be incorporated into initial conditions; it is thus resonable to apply the Liapunov stability concept i.e. stability with respect to initial conditions.

4.1 First Model in Chemical Kinetics

The results in what follows are essentially based on a paper by Kružkov and Peregudov (1990). We shall study the properties of the system

$$x' = F(x) \tag{4.1.1}$$

under following assumptions:

i) $F : Q \subset R^N \longrightarrow R^N$, $Q = \{x \in R^N, x^i \geq 0, i = 1, \ldots, N\}$;

ii) $F(0) = 0$;

iii) $\frac{\partial F^i}{\partial x^j}(x) \geq 0$, $i \neq j$;

iv) $\sum_1^N F^i(x) \equiv 0$.

These assumptions are motivated by properties in chemical kinetics (see the book of Frank – Kamenetskii, 1987).

Proposition 4.1 *If* $x(\cdot)$, $y(\cdot)$ *are solutions of* (4.1.1) *and* $x^i(0) \geq y^i(0)$ *for all* i, *then* $x^i(t) \geq y^i(t)$ *for all* i *and* $0 \leq t \leq \hat{t} \in I_x \cap I_y$, I_x *and* I_y *being the intervals of definition for solutions* $x(\cdot)$, $y(\cdot)$ *respectively.*

Proof Let $z(t) = x(t) - y(t)$; then

$$\frac{d}{dt} z^i(t) = F^i(x(t)) - F^i(y(t)) = \sum_{j=1}^{N} \frac{\partial F^i}{\partial x^j}(\xi(t))[x^j(t) - y^j(t)] =$$

$$= \sum_{j=1}^{N} \frac{\partial F^i}{\partial x^j}(\xi(t)) z^j(t)$$

Define $w(t) = z(t)e^{\lambda t}$; then

$$\frac{d}{dt} w^i(t) = e^{\lambda t} \frac{d}{dt} z^i(t) + \lambda e^{\lambda t} z^i(t) =$$

$$= \sum_{j=1}^{N} \frac{\partial F^i}{\partial x^j}(\xi(t)) w^j(t) + \lambda w^i(t)$$

If $\lambda > 0$ is large enough, all coefficients are positive and the unique solution $w(\cdot)$, constructed by successive approximations

$$w_{k+1}^i(t) = w^i(0) + \int_0^t \sum_{j=1}^{N} \left[\frac{\partial F^i}{\partial x^j}(\xi(\tau)) + \lambda \delta_j^i \right] w_k^j(\tau) d\tau,$$

$w^i(t) = \lim_{k \to \infty} w_k^i(t)$ satisfies $w^i(t) \geq 0$, hence, $z^i(t) \geq 0$ that means $x^i(t) \geq y^i(t)$, $0 \leq t \leq \hat{t}$.

Corollary 4.1 *If* $x^i(0) \geq 0$ *for all* i *then* $x^i(t) \geq 0$ *for all* i *and* $0 \leq t \leq \hat{t}$. *(We compare with* $y \equiv 0$*).*

Corollary 4.2 *All solutions are bounded, hence globally defined.*

Proof From $(x^i)'(t) = F^i(x(t))$ we deduce $\sum_1^N (x^i)'(t) = \sum_1^N F^i(x(t)) \equiv 0$ hence, $t \longmapsto \sum_1^N (x^i)(t)$ is constant, $0 \leq \sum_1^N x^i(t) = \sum_1^N x^i(0)$, hence, for all j, $0 \leq x^j(t) \leq \sum_1^N x^i(0)$.

Proposition 4.2 *Denote* $||x|| = \sum_1^N |x^i|$. *If* x, y *are solutions to the system, then* $||x(t) - y(t)|| \leq ||x(t_0) - y(t_0)||$ *for all* $t \geq t_0$.

Proof Let $z^i(t_0) = \max\{x^i(t_0), y^i(t_0)\}$. We have $z^i(t_0) = x^i(t_0)$ for $i \in I_1$ $(x^i(t_0) \geq y^i(t_0))$ and $z^i(t_0) = y^i(t_0)$ for $i \in I_2$ $(x^i(t_0) \leq y^i(t_0))$. From here

$$||z(t_0) - x(t_0)|| + ||z(t_0) - y(t_0)|| = \sum_1^N |z^i(t_0) - x^i(t_0)| +$$

$$+ \sum_1^N |z^i(t_0) - y^i(t_0)| = \sum_{i \in I_1} |x^i(t_0) - y^i(t_0)| +$$

$$+ \sum_{i \in I_2} |y^i(t_0) - x^i(t_0)| = ||x(t_0) - y(t_0)||.$$

We have also $x^i(t_0) \leq z^i(t_0)$, $y^i(t_0) \leq z^i(t_0)$ for all i, hence

$$||z(t) - x(t)|| = \sum_1^N [z^i(t) - x^i(t)] = \sum_1^N [z^i(t_0) - x^i(t_0)] =$$

$$= ||z(t_0) - x(t_0)||$$

and in the same way

$$||z(t) - y(t)|| = ||z(t_0) - y(t_0)||.$$

We deduce that

$$||x(t) - y(t)|| \leq ||x(t) - z(t)|| + ||z(t) - y(t)|| =$$

$$= ||z(t_0) - x(t_0)|| + ||z(t_0) - y(t_0)|| = ||x(t_0) - y(t_0)||.$$

Corollary 4.3 *If* x *is a solution of the system and* $\xi(t) = ||x'(t)||$, *then* $\xi(t)$ *is decreasing.*

Proof Let $x(\cdot)$ be a solution to the system and $y(\cdot)$ defined by $y(t) = x(t+h)$; $y(\cdot)$ is also a solution and from Proposition 4.2 we deduce $||x(t+h) - x(t)|| \leq ||x(\tau+h) - x(\tau)||$ for $t \geq \tau$, hence

$$\left\|\frac{x(t+h)-x(t)}{h}\right\| \leq \left\|\frac{x(\tau+h)-x(\tau)}{h}\right\|$$

and for $h \to 0$ we obtain $||x'(t)|| \leq ||x'(\tau)||$.

Remark 4.1 *The function* V *defined by* $V(x) = \sum_1^N |F^i(x)| = ||F(x)||$ *is a Liapunov function, since for all solutions* $x(\cdot)$, $t \longmapsto V(x(t)) = ||F(x(t))|| = ||x'(t)||$ *is decreasing.*

Remark 4.2 *Let* $x(\cdot)$ *be an arbitrary solution,* $f^i(t) = F^i(x(t))$; *we have* $\sum_1^N f^i(t) \equiv 0$, $(f^i)'(t) = \sum_{j=1}^N \frac{\partial F^i}{\partial x^j}(x(t))f^j(t)$; *let* $\tau \geq 0$, $f^i_+(\tau) = \max(0, f^i(\tau))$, $f^i_-(\tau) = \min(0, f^i(\tau))$. *We have* $f^i_+(\tau) \geq 0$ *for all* i, $f^i_-(\tau) \leq 0$ *for all* i *and since* f^i *are solutions for a linear system with positive nondiagonal elements, we deduce that* $f^i_+(t) \geq 0$, $f^i_-(t) \leq 0$ *for all* $t \geq \tau$.

From $f^i(\tau) = f^i_+(\tau) + f^i_-(\tau)$ *it follows that* $f^i(t) = f^i_+(t) + f^i_-(t)$ *and* $\sum_1^N f^i_+(t) = \sum_1^N f^i_+(\tau)$, $\sum_1^N f^i_-(t) = \sum_1^N f^i_-(\tau)$; *the last relations follow from the fact that if*

$$(z^i)'(t) = \sum_1^N \frac{\partial F^i}{\partial x^j}(\xi(t))z^j(t)$$

then

$$\left[\sum_1^N z^i\right]'(t) = \sum_1^N (z^i)'(t) = \sum_{i=1}^N \sum_{j=1}^N \frac{\partial F^i}{\partial x^j}(\xi(t))z^j(t) = 0$$

hence $t \longmapsto \sum_1^N z^i(t)$ *is constant. We have*

$$(f^i_+)'(t) = \sum_1^N \frac{\partial F^i}{\partial x^j}(\xi(t))f^j_+(t) \geq \frac{\partial F^i}{\partial x^i}(\xi(t))f^i_+(t)$$

hence

$$f^i_+(t) \geq f^i_+(\tau) \cdot \exp\left(\int_\tau^t \frac{\partial F^i}{\partial x^i}(\xi(s))ds\right)$$

for $t \geq \tau$. *In the same way*

$$f^i_-(t) \leq f^i_-(\tau) \cdot \exp\left(\int_\tau^t \frac{\partial F^i}{\partial x^i}(\xi(s))ds\right)$$

for $t \geq \tau$. *We deduce that either* $f^i_+(t) \equiv 0$ *for all* $t \geq t_0$ *or* $f^i_+(t) > 0$ *for all* $t \geq t_0$ *and also either* $f^i_-(t) \equiv 0$ *or* $f^i_-(t) < 0$ *for all* $t \geq t_0$. *(If* $f^i_+(t) = 0$ *then* $f^i_+(\tau) = 0$ *for all* $\tau \leq t$, *and if* $f^i_-(t) = 0$ *then* $f^i_-(\tau) = 0$ *for all* $\tau \leq t$*).*

Lemma 4.1 *If there exists* τ_0 *such that* $V(x(\tau_0)) > 0$ *for* V *as in Remark* 4.2, *then there exists* $\tau_1 > \tau_0$ *such that* $V(x(\tau_1)) < V(x(\tau_0))$, *that is if* $t \longmapsto V(x(t))$ *is not identically zero, it cannot be constant.*

Proof According to the facts described in Remark 4.2, there exists $I_1 \subset (1, N)$ such that $f^i_+(t) > 0$ for $i \in I_1$, $f^i_+(t) \equiv 0$ for $i \notin I_1$, $t \geq \hat{\tau} > \tau_0$. If $f^i_-(t) \equiv 0$ for all $i \in I_1$ we have

$$V(x(t)) = \sum_1^N |f^i(t)| = \sum_1^N |f^i_+(t) + f^i_-(t)| =$$

$$= \sum_{i \in I_1} f^i_+(t) + \sum_{i \notin I_1} |f^i_-(t)| = \sum_{i \in I_1} f^i_+(t) - \sum_{i \notin I_1} f^i_-(t)$$

But $\sum_1^N f^i(t) = \sum_1^N f^i_+(t) + \sum_1^N f^i_-(t) \equiv 0$ hence $\sum_{i \in I_1} f^i_+(t) + \sum_{i \notin I_1} f^i_-(t) \equiv 0$ and

$$V(x(t)) = 2 \sum_{i \in I_1} f^i_+(t) = 2 \sum_{i \in I_1} f^i(t) =$$

$$= 2 \sum_{i \in I_1} (x^i)'(t) = 2 \left(\sum_{i \in I_1} x^i(t)\right)'$$

If $t \longmapsto V(x(t))$ is constant for $t \geq \tau_0$ then $t \longmapsto \sum_{i\in I_1} x^i(t)$ is polynomial of degree one, unless $V(x(t)) \equiv 0$ and this contradicts boundedness of $t \longmapsto \sum_{i\in I_1} x^i(t)$; so that Lemma is proved if $f^i_-(t) \equiv 0$ for all $i \in I_1$. If there exists $i \in I_1$ and $\tau_1 > \hat{\tau}$ with $f^i_-(\tau_1) < 0$ we have

$$V(x(\tau_1)) = \sum_1^N |f^j_+(\tau_1) + f^j_-(\tau_1)| \leq \sum_{j\neq i}(|f^j_+(\tau_1)| + |f^j_-(\tau_1)|) +$$

$$+|f^i_+(\tau_1)| - |f^i_-(\tau_1)| < \sum_{j=1}^N(|f^j_+(\tau_1)| + |f^j_-(\tau_1)|) =$$

$$= \sum_1^N(|f^j_+(\tau_0)| + |f^j_-(\tau_0)|) = \sum_1^N |f^j(\tau_0)| = V(x(\tau_0)).$$

(Let us note that $f^j_+(\tau_0) = |f^j_+(\tau_0)| = |f^j(\tau_0)|$ if $f^j(\tau_0) > 0$ and $-f^j_-(\tau_0) = |f^j_-(\tau_0)| = |f^j(\tau_0)|$ if $f^j(\tau_0) < 0$).

We are now in position to state and prove

Theorem 4.1 *For every* $M > 0$ *there exist equilibria* $\hat{x}$ *such that* $\sum_1^N \hat{x}^i = M$, *and every solution* $x(\cdot)$ *with* $\sum_1^N x^i(0) = M$ *tends to such equilibrium.*

Proof Let $x(\cdot)$ be a solution with $\sum_1^N x^i(0) = M$; then, as in Corollary 4.2 we have $\sum_1^N x^i(t) = M$ and the solution is bounded. It follows existence of a sequence $(t_m)_m$ with $\lim_{m\to\infty} t_m = \infty$ and $\lim_{m\to\infty} x(t_m) = \hat{x}$.

The function V is continuous, hence $\lim_{m\to\infty} V(x(t_m)) = V(\hat{x})$. On the other hand $t \longmapsto V(x(t))$ is not increasing, hence $\lim_{t\to\infty} V(x(t)) = V(\hat{x})$.

Consider the solution $\tilde{x}(\cdot)$ with $\tilde{x}(0) = \hat{x}$; since $\hat{x}$ is an ω–limit point for the solution $x(\cdot)$, it follows that for all $t, \tilde{x}(t)$ is an ω–limit point for the solution $x(\cdot)$ (recall the fact that the ω–limit set is invariant); as above we deduce that $V(\tilde{x}(t)) = V(\hat{x})$, hence $t \longmapsto V(\tilde{x}(t))$ is constant. According to Lemma 4.1, we deduce that $V(\hat{x}) = 0$, hence $\lim_{t\to\infty} V(x(t)) = 0$, that is $\lim_{t\to\infty} \sum_1^N |F^i(x(t))| = 0$, hence $\lim_{t\to\infty} F^i(x(t)) = 0$. We deduce that

$$\lim_{m\to\infty} F^i(x(t_m)) = F^i(\lim_{t\to\infty}(x(t_m)) = F^i(\hat{x}) = 0$$

hence $\hat{x}$ is an equilibrium. Since $t \longmapsto \sum_1^N x^i(t)$ is constant, it follows that $\sum_1^N x^i(t_m) = \sum_1^N x^i(0) = M$; hence, $\lim_{m\to\infty} \sum_1^N x^i(t_m) = \sum_1^N \hat{x}^i = M$. We have further $||x(t) - \hat{x}|| \leq ||x(t_m) - \hat{x}||$ for $t \geq t_m$ (see Proposition 2) and from $\lim_{m\to\infty} ||x(t_m) - \hat{x}|| = 0$ we deduce that $\lim_{t\to\infty} ||x(t) - \hat{x}|| = 0$. The theorem is proved.

Remark 4.3 *It can be proved that if* $\frac{\partial F^i}{\partial x^j}(x) > 0$ *for* $i \neq j$ *and* $x^j > 0$, *then for given* M *there exists a unique* $\hat{x}$ *with* $\sum_1^N \hat{x}^i = M$ *and* $F^i(\hat{x}) = 0$ *for all* i. *In this situation, the unique equilibrium* $\hat{x}$ *is globally asymptotically stable on the invariant set defined by* $\sum_1^N x^i = M$.

4.2 Stability of Closed Chemical System Subject to Mass – Action Law

The Mathematical Model

A basic concept in chemical kinetics is the concept of reaction velocity. Given the substances A_i which are reactants in some chemical reaction and the substances B_i which are products of the same reaction, the stoichiometric coefficients α_i of the reactants and β_i of the reaction products (both nonnegative integers), the reaction velocities are defined as $(-\frac{1}{V})(\frac{dN_i}{dt})$ for reactants and $(\frac{1}{V})(\frac{dN_i'}{dt})$ for products, the minus indicating that the reactant quantity is diminishing during the reaction. Here V denotes the volume of the reaction enclosure and N_i, N_i' denote the current quantities of reactant A_i and product B_i expressed by some unit of measurement (e.g. in moles).

It is assumed in physical chemistry that "between various substance quantities occurring in some time interval a stoichiometric proportion is valid" (I.G.Murgulescu et al., 1981).

This is a definition of a unique reaction velocity for the given reaction. This velocity is sometimes called reaction advance:

$$-\frac{1}{V}\frac{1}{\alpha_1}\frac{dN_1}{dt} = -\frac{1}{V}\frac{1}{\alpha_2}\frac{dN_2}{dt} = \ldots = \frac{1}{V}\frac{1}{\beta_1}\frac{dN_1'}{dt} = \ldots = w$$

Introducing the volumetric concentrations $c_i = N_i/V$, for any reactant one can obtain a differential equation of the form

$$\dot{c}_i = -\alpha_i w$$

and for any product a differential equation of the form

$$\dot{c}_i' = \beta_i w.$$

For the so-called reversible reactions, the same substance can be both reactant and product; therefore two stoichiometric coefficients are associated to this substance: the reactant coefficient α_i and the product coefficient β_i (if some substance is a reactant or a product only then either $\beta_i = 0$ or $\alpha_i = 0$). Employing the convention that the plus corresponds to product generation and the minus to reactant consumption respectively, it is natural to assume – according to the practical experiments – that the speed of generating some substance is the difference of the generating speeds in both senses:

$$\frac{dc_i}{dt} = \frac{dc_i^+}{dt} - \frac{dc_i^-}{dt}.$$

But the stoichiometric proportion is still valid for each case separately. Therefore

$$-\frac{1}{\alpha_1}\frac{dc_i^-}{dt} = \ldots = \frac{1}{\beta_1}\frac{dc_i^+}{dt} = \ldots = w$$

and, from here

$$\dot{c}_i = (\beta_i - \alpha_i)w.$$

For the case of several simultaneous reactions (when the substances react in several stages) the above considerations are still true for each reaction (stage) taken separately. Under such circumstances, with some reactant A_i one associates a stoichiometric coefficient α_{ij}, corresponding to each reaction j and with some product B_k we associte a stoichiometric coefficient β_{kj} which corresponding to each reaction j; for each reaction j the reaction velocity (advance) w_j is defined. But the amount of substance generated in the reaction enclosure per time unit is a sum of the amounts generated per time unit in each stage, hence

$$\dot{c}_i = -\sum_j \alpha_{ij} w_j$$

for reactants and

$$\dot{c}'_i = \sum_j \beta_{ij} w_j$$

for products. For reversible multistage reactions the same considerations will give

$$\dot{c}_i = \sum_j (\beta_{ij} - \alpha_{ij}) w_j.$$

The expressions for reaction velocities are established on the basis of the *mass action law* (op.cit., 1981): it is assumed that at the molecular level only collision implies reaction. It follows that the reaction velocity is proportional to the amounts of reacting substances. With some additional empirical consideration one gets for a given reaction

$$w = k c_1^{\nu_1} c_2^{\nu_2} \ldots (c'_k)^{\nu'_k}$$

where the real numbers ν_j, ν'_k are called the reaction orders of the corresponding substances, and k is called activating constant; the activating constant depends on the thermodynamic parameters mainly on the temperature.

If the thermodynamic state is close to the thermodynamic equilibrium, the reaction order equals the stoichiometric coefficient of the substance: $\nu_i = \alpha_i$, $\nu'_k = \beta_k$. For reversible reactions $w = w^+ - w^-$ i.e.

the reaction velocity is the difference of the velocities in the two senses; here w^+ and w^- have the usual form

$$w^+ = k^+ \prod c_i^{\alpha_i}, \qquad w^- = k^- \prod c_i^{\beta_i}$$

All these considerations, which can be found in basic chemical kinetics work (e.g., op.cit., 1981) lead to the model which will be tackled below.

Consider a list of m substances A_i, $i = 1, \dots, m$, which enter in n reversible reactions. Denote by α_{ij} the stoichiometric coefficient of A_i as reactant in the reaction j and by β_{ik} the stoichiometric coefficient of the same A_i as product in the reaction k. By the so-called stoichiometric equations written in the following symbolic form

$$\sum_{i=1}^{m} \alpha_{ij} A_i \rightleftarrows \sum_{i=1}^{m} \beta_{ij} A_i\,, \quad j = 1, \dots, n \tag{4.2.1}$$

the so-called *reaction mechanism* is defined.

Assuming the chemical system to be closed (i.e., no substance exchange with the environment takes place), the thermodynamic parameters to be constant, and neglecting the diffusion, the following equations for concentrations dynamics are obtained:

$$\dot{c}_i = \sum_{j=1}^{n} (\beta_{ij} - \alpha_{ij}) w_j(c) \tag{4.2.2}$$

where $w_j(c) = w_j^+(c) - w_j^-(c)$ and $w_j^{\pm}(c)$ are supposed to obey the formal kinetics formulae deduced from the mass action law

$$w_j^+(c) = k_j^+ \prod_{i=1}^{m} (c_i)^{\alpha_{ij}}, \qquad w_j^-(c) = k_j^- \prod_{i=1}^{m} (c_i)^{\beta_{ij}} \tag{4.2.3}$$

where some stage is more or less reversible according to the ratio of the activating coefficients k_j^+ and k_j^-.

In the following some assumptions deduced from physical considerations are given and some properties validating the model are established on this basis.

The Coefficients Hypothesis. The coefficients α_{ij} and β_{ij} are nonnegative integers with the property that for any j there exists i such that $\alpha_{ij} + \beta_{ij} \neq 0$. Indeed, α_{ij} and β_{ij} are nonnegative integers because they are stoichiometric coefficients indicating the number of molecules of reactant (or product) participating in some reaction. The above relation shows that at least one substance must participate in each reaction (as reactant or as product) otherwise, that reaction does not exist in the considered chemical system.

Concentration Nonnegativeness. Obviously, the concentration of any of substance cannot be negative. This system property is valid at any moment of time and, mathematically speaking, it is a property of the solutions of the system of differential equations: it cannot be postulated as the first hypothesis – a simple coefficient property. In this case the nonnegativeness of the initial conditions should imply the nonnegativeness of the solutions for all $t > 0$ i.e. the system should have an invariant set. The following result can be formulated:

Proposition 4.3 *Consider system* (4.2.2) *whose coefficients are such that for any* j *there exists at least one* i *with the property that* $\alpha_{ij} + \beta_{ij} \neq 0$. *If* $c_i(0) \geq 0$, $i = 1, \dots, m$, *then for any* i *either* $c_i(t) > 0$ *or* $c_i(t) \equiv 0$ *on the whole definition set of the solution.*

Proof Assume $c_i(0) \geq 0$ to be given and let (T_1, T_2) be the maximal existence interval for the corresponding solutions (the RHS of the system is polynomial, hence the existence of solutions on finite intervals follows from any existence theorem).

Consider the equation

$$\dot{c}_k = \sum_{j=1}^{n} (\beta_{kj} - \alpha_{kj})(w_j^+(c) - w_j^-(c))$$

and re-write it as follows

$$\dot{c}_k = -\sum_{j=1}^{n} (\alpha_{kj} w_j^+(c) + \beta_{kj} w_j^-(c)) +$$

$$+\sum_{j=1}^{n}(\beta_{kj}w_j^+(c)+\alpha_{kj}w_j^-(c)).$$

Here $\alpha_{kj} \geq 0$, $\beta_{kj} \geq 0$, but since these are stoichiometric coefficients i.e. nonnegative integers, $\alpha_{kj} \neq 0$ implies $\alpha_{kj} \geq 1$ and the same is valid for β_{kj}. From the Coefficients Hypothesis, there exists some j such that $\alpha_{kj}+\beta_{kj} \neq 0$ i.e. either a α_{kj} or a β_{kj} is nonzero. Then, taking into account the form of w_j^+ and w_j^-, it follows

$$\dot{c}_k = -a_k(t)c_k + b_k(t)$$

where $a_k(t)$ and $b_k(t)$ are continuous on (T_1, T_2).

Therefore

$$c_k(t) = c_k(0)\exp\left(-\int_0^t a_k(\tau)d\tau\right) +$$

$$+\int_0^t \exp\left(-\int_\tau^t a_k(\theta)d\theta\right) b_k(\tau)d.\tau \qquad (4.2.4)$$

Consider $\epsilon > 0$ and the solution $c^\epsilon(t)$ with the initial condition $c_k(0)+\epsilon$ for all k. From the continuity with respect to the initial conditions it follows that $c_k^\epsilon(t)$ is defined on any compact set contained in (T_1, T_2) provided $\epsilon > 0$ is sufficiently small. Consider the set $\mathcal{M}_\epsilon = \{t \,|\, c_k^\epsilon(\tau) > 0,\ 0 \leq \tau < t,\ \forall k\}$ and let $\mu_\epsilon = \sup \mathcal{M}_\epsilon$. From (4.2.4), we see that $c_k^\epsilon(\mu_\epsilon) > 0$ for all k because $b_k(t) > 0$ on $\mathcal{M}_\epsilon$, hence $c_k^\epsilon(t) > 0$ for $t > \mu_\epsilon$ close enough to μ_ϵ, what contradicts the maximality of μ_ϵ. It follows that $c_k^\epsilon(t) > 0$ for any $t \in (T_1^\epsilon, T_2^\epsilon)$, hence for any compact set contained in (T_1, T_2); for $t \in (T_1, T_2)$ pick a compact set containing t and let $\epsilon \to 0$. It follows that $c_k(t) \geq 0$ for any $t \in (T_1, T_2)$ and all k. Consider now a $\hat{t}$ such that $c_k(\hat{t}) = 0$; from (4.2.4) it follows that $c_k(0) = 0$, $b_k(\tau) \equiv 0$, $0 \leq \tau \leq \hat{t}$ hence $c_k(t) = 0$ for $0 \leq t \leq \hat{t}$. Since the solutions of the system are analytic, it follows that $c_k(t) \equiv 0$ on the whole definition interval, what ends the proof.

We note that existence of the invariant set for the system of differential equations, describing a property of the physical system, namely, the concentrations nonnegativeness, represents a justification of the model based on mass action law. For other situations the proof has to be redone.

Definition 4.1 *Any point with positive coordinates c_k, $k = 1, \ldots, m$ is called an* <u>*admissible point*</u>*. The set of the admissible points $\{c \mid c_k > 0,\ k = 1, \ldots, m\} \subset R^m$ is called* <u>*admissible set*</u>*.*

The first integral and the invariant hyperplane. We introduce the matrix of stoichiometric coefficients:

$$G = \begin{pmatrix} \beta_{11} - \alpha_{11} & \beta_{12} - \alpha_{12} & \ldots & \beta_{1n} - \alpha_{1n} \\ \beta_{21} - \alpha_{21} & \beta_{22} - \alpha_{22} & \ldots & \beta_{2n} - \alpha_{2n} \\ \vdots & \vdots & & \vdots \\ \beta_{m1} - \alpha_{m1} & \beta_{m2} - \alpha_{m2} & \ldots & \beta_{mn} - \alpha_{mn} \end{pmatrix}$$

and the vector of the reaction velocities $w = \mathrm{col}(w_1, \ldots, w_n)$. Consequently system (4.2.2) can be written as follows

$$\dot{c} = Gw(c). \tag{4.2.5}$$

Denote $r = \mathrm{rank}\, G$ $(0 \le r \le \min(m, n))$; by renumbering the substances and the reactions, the following partition of G is obtained

$$G = \begin{pmatrix} G_{11} & G_{12} \\ G_{21} & G_{22} \end{pmatrix}$$

where the $r \times r$ matrix G_{11} is nonsingular. Denote by (c^r, c^{m-r}), (w^r, w^{n-r}) the corresponding partitions of c and w. With this, system (4.2.5) can be written as follows

$$\dot{c}^r = G_{11} w^r(c) + G_{12} w^{n-r}(c)$$

$$\dot{c}^{m-r} = G_{21} w^r(c) + G_{22} w^{n-r}(c). \tag{4.2.6}$$

Since the rank $G = r$ and G_{11} is nonsingular, the last $n - r$ columns of G are linear combinations of the first r columns. Therefore, a matrix H of corresponding dimensions exists such that

$$G_{12} = G_{11}H, \qquad G_{22} = G_{21}H.$$

It follows that

$$H = G_{11}^{-1}G_{12}, \qquad G_{22} = G_{21}G_{11}^{-1}G_{12}$$

and, substituting into (4.2.6), we find:

$$\dot{c}^{m-r} = G_{21}G_{11}^{-1}\left[G_{11}w^{r}(c) + G_{12}w^{n-r}(c)\right] = G_{21}G_{11}^{-1}\dot{c}^{r}.$$

Therefore

$$\frac{d}{dt}\left(c^{m-r} - G_{21}G_{11}^{-1}c^{r}\right) = 0$$

what defines $c^{m-r} - G_{21}G_{11}^{-1}c^{r}$ as a (vector) first integral of the system. In fact any solution of the system belongs to the linear invariant manifold

$$\mathcal{L}(c) \equiv c^{m-r} - G_{21}G_{11}^{-1}c^{r} = c^{m-r}(0) - G_{21}G_{11}^{-1}c^{r}(0) \qquad (4.2.7)$$

called "substance balance plane" (although it is in fact a hyperplane).

Equilibrium Points

As it has been already mentioned, the stability property is not, generally speaking, a property of the system, but a property of a certain solution. Usually stability of some remarkable solutions is studied; among them the most frequent case is that of the stationary solutions. This is exactly the case of the considered chemical system whose stationary solutions are given by the singular points of the system (4.2.5), namely by the solutions of the system

$$Gw(c) = 0. \qquad (4.2.8)$$

In chemical kinetics not all solutions of system (4.2.8) are of interest but only the so-called *detailed balance* points which are equilibrium points of each reaction (stage) considered separately.

Definition 4.2 *A singular point of system (4.2.5) satisfying the algebraic system*

$$w_j^+(c) = w_j^-(c), \quad j = 1, \ldots, n \tag{4.2.9}$$

is called a detailed balance point.

It is obvious that the detailed balance points are indeed singular points because (4.2.9) implies $w = 0$ hence (4.2.8) is verified. Less obvious is the converse: is it true that the detailed balance points are the only singular points? For instance, if $r = \operatorname{rank} G = n$, this holds because (4.2.8) implies $w = 0$. The equality $r = n$ is not impossible for real chemical systems because $r \leq \min(n, m)$ and the Gibbs phase rule (op.cit., 1981) shows that $n \leq m$. However, if $r \neq n$ the relation between the set of all equilibrium points and the set of detailed balance points is no longer straightforward.

Still a remark must be made: among the detailed balance points only those belonging to the admissible set $\{c : c_k > 0, \ k = 1, \ldots, m\}$ are of interest.

Definition 4.3 *A detailed balance point belonging to the admissible set is called admissible detailed balance point.*

In other words, only those detailed balance points which are also admissible points are of interest. Determination of the admissible detailed balance points requires finding of the positive solutions for the system

$$k_j^+ \prod_{i=1}^{m} (c_i)^{\alpha_{ij}} = k_j^- \prod_{i=1}^{m} (c_i)^{\beta_{ij}}, \quad j = 1, \ldots, m.$$

But $c_i > 0$; it follows

$$\sum_{i=1}^{m} (\beta_{ij} - \alpha_{ij}) \ln c_i = \ln(k_j^+ / k_j^-).$$

Denoting $a_i = \ln c_i$, $b = \mathrm{col}(\ln(k_1^+/k_1^-), \ldots, \ln(k_n^+/k_n^-))$, we obtain the following linear nonhomogenous system

$$G^* a = b. \tag{4.2.10}$$

If the same partition of G as in the case of first integral is used, then (4.2.10) can be written as follows

$$G_{11}^* a^r + G_{21}^* a^{m-r} = b^r$$

$$G_{12}^* a^r + G_{22}^* a^{n-r} = b^{n-r}. \tag{4.2.11}$$

If the expression for G_{22} is taken into account, the following necessary and sufficient condition of compatibility for (4.2.10) arises

$$b^{n-r} = G_{12}^* H (G_{11}^*)^{-1} b^r. \tag{4.2.12}$$

For given matrix of stoichiometric coefficients condition, (4.2.12) is a restriction put on the constants $k_j^\pm$; for instance, admissible detailed balance points can exist for a certain range of temperatures only.

Due to the fact that the system has a first integral, its solutions belong to an invariant linear manifold (4.2.7). An interesting problem is to know if admissible detailed balance points can be found in this manifold. The answer is given by the following result of Ya.B.Zeldovič (1938, see also V.M.Vasiliev et al., 1973).

Theorem 4.2 *If system* (4.2.5) *has an admissible detailed balance point and in the linear manifold* $\mathcal{L}(c) = q$ *there exists an admissible point (with positive coordinates), then in this manifold there exists a unique admissible detailed balance point.*

Proof Consider the invariant manifold

$$c^{m-r} - G_{21} G_{11}^{-1} c^r = q$$

and let $\hat{c}$ be an admissible point of this manifold; therefore $\hat{c}_i > 0$, $i = 1, \ldots, m$ and $\hat{c}^{m-r} - G_{21} G_{11}^{-1} \hat{c}^r = q$. According to the assumptions, there exist admissible detailed balance points, hence the system (4.2.10) is compatible and reduces to

$$a^{r} = (G_{11}^{*})^{-1} b^{r} - (G_{11}^{*})^{-1} G_{21}^{*} a^{m-r}. \tag{4.2.13}$$

On the other hand, the coordinates of any admissible point (being positive) can be written as $c_i = e^{a_i}$; it follows that the admissible points belonging to the invariant hyperplane verify the system

$$e^{a_i} - \sum_{k=1}^{r} \chi_{ik} e^{a_k} = \mu_i, \quad i = r+1, \ldots, m \tag{4.2.14}$$

where χ_{ik} are the elements of the matrix $H = G_{21} G_{11}^{-1}$ and μ_i are the components of q.

By assumption, each of the systems (4.2.13) and (4.2.14) has at least one solution. The theorem states that there exists a *unique* solution verifying both systems simultaneously. In order to prove this, consider the function $\Gamma_{\hat{c}} : R^m \longrightarrow R$ defined by

$$\Gamma_{\hat{c}}(a) = \sum_{1}^{m} \left(e^{a_i} - a_i e^{\hat{a}_i} \right)$$

where $\hat{a}_i = \ln \hat{c}_i$. The restriction of Γ (the index $\hat{c}$ was introduced when defining the function to stress its dependence on the admissible point from the invariant manifold, will be suppressed from now on) to the linear manifold (4.2.13) is $\tilde{\Gamma} : R^{m-r} \longrightarrow R$ defined by

$$\tilde{\Gamma}(a^{m-r}) = \sum_{i=1}^{r} \left[\exp\left(\beta_i - \sum_{k=r+1}^{m} \chi_{ki} a_k \right) - \right.$$

$$\left. - e^{\hat{a}_i} \left(\beta_i - \sum_{k=r+1}^{m} \chi_{ki} a_k \right) \right] + \sum_{i=r+1}^{m} \left(e^{a_i} - a_i e^{\hat{a}_i} \right) \tag{4.2.15}$$

where β_i are the components of the vector $(G_{11}^{*})^{-1} b^{r}$. It is stated that any critical point of $\tilde{\Gamma}$ is the R^{m-r}–projection of a common solution of systems (4.2.13), (4.2.14) and vice – versa.

Indeed, the first order partial derivatives of $\tilde{\Gamma}$ are

$$\frac{\partial \tilde{\Gamma}}{\partial a_l}(a^{m-r}) = \sum_{i=1}^{r} \left[(-\chi_{li}) \exp\left(\beta_i - \sum_{k=r+1}^{m} \chi_{ki} a_k \right) + \chi_{li} e^{\hat{a}_i} \right] +$$

$$+ e^{a_l} - e^{\hat{a}_l}, \quad l = r+1, \ldots, m$$

But $\hat{a}_i$ satisfy (4.2.14), hence

$$\frac{\partial \tilde{\Gamma}}{\partial a_l}(a^{m-r}) = e^{a_l} - \sum_{i=1}^{r} \chi_{li} \exp\left(\beta_i - \sum_{k=r+1}^{m} \chi_{ki} a_k \right) -$$

$$- \mu_l, \quad l = r+1, \ldots, m.$$

Let $(a_{r+1}, \ldots, a_m)$ be a critical point and define a_j, $j = 1, \ldots, r$, by (4.2.13). Then a_i, $i = 1, \ldots, m$ satisfy simultaneously (4.2.13), (4.2.14), hence the coordinates of the critical point represent the corresponding coordinates of a common solution of (4.2.13), (4.2.14). Conversely, let a_i, $i = 1, \ldots, m$, be a solution of (4.2.13), (4.2.14). Because (4.2.13), (4.2.14) are satisfied, $\dfrac{\partial \tilde{\Gamma}}{\partial a_l}$ are given by

$$\frac{\partial \tilde{\Gamma}}{\partial a_l}(a^{m-r}) = e^{a_l} - \sum_{i=1}^{r} \chi_{li} e^{a_i} - \mu_l = 0$$

hence the last coordinates correspond to a critical point.

On the other hand, $\tilde{\Gamma}$ is defined everywhere in R^{m-r} and for $|a^{m-r}| \to \infty$, $\tilde{\Gamma} \to \infty$. Indeed, elementary arguments show that $\varphi(\xi) = e^{\xi} - \beta\xi$, $\xi > 0$, is bounded from below by $\beta(1 - \ln\beta)$. It follows then from (4.2.15) that

$$\tilde{\Gamma}(a^{m-r}) \geq \sum_{i=r+1}^{m} \left(e^{a_i} - e^{\hat{a}_i} \right) + \sum_{i=1}^{r} e^{\hat{a}_i}(1 - \hat{a}_i)$$

and the above formulated property follows. (In fact each term of the first sum is bounded from below and if only one component tends to infinity the function tends to infinity as well). Therefore, the absolute minimum of $\tilde{\Gamma}$ exists, hence a critical point *exists*. This critical point is unique. Indeed, the Hessian matrix of second order partial derivatives has the components

$$\frac{\partial \tilde{\Gamma}}{\partial a_k \partial a_l}(a^{m-r}) = e^{a_l}\delta_{kl} + \sum_{i=1}^{r} \chi_{li} \exp\left(\beta_i - \sum_{j=r+1}^{m} \chi_{ji} a_j\right) \chi_{ki}$$

where $\delta_{ij} = 0$, $i \neq j$, $\delta_{ii} = 1$. Introducing the diagonal matrices

$$A^{m-r} = \operatorname{diag}(e^{a_{r+1}}, \dots, e^{a_m})$$

$$A^r = \operatorname{diag}\left(\exp\left(\beta_1 - \sum_{r+1}^{m} \chi_{j1} a_j\right), \dots, \exp\left(\beta_r - \sum_{r+1}^{m} \chi_{jr} a_j\right)\right)$$

and taking into account that χ_{li} are components of the matrix $G_{21}G_{11}^{-1}$, it follows

$$\mathcal{H}(\tilde{\Gamma}(a^{m-r})) = A^{m-r} + G_{21}G_{11}^{-1}A^r(G_{11}^*)^{-1}G_{21}^* > 0.$$

Therefore, if we take the Taylor expansion, we get for any critical point $\tilde{a}^{m-r}$

$$\tilde{\Gamma}(a^{m-r}) - \tilde{\Gamma}(\tilde{a}^{m-r}) =$$

$$= (a^{m-r} - \tilde{a}^{m-r})^* \mathcal{H}(\tilde{\Gamma}(P^{m-r}))(a^{m-r} - \tilde{a}^{m-r}) > 0$$

if $a^{m-r} \neq \tilde{a}^{m-r}$, hence any critical point is an isolated minimum point. On this basis, the uniqueness of the minimum point follows. Indeed, if $\bar{a}^{m-r}$ is the absolute minimum point whose existence has been just proved and if $\tilde{a}^{m-r}$ is another minimum point the above inequality gives

$$\tilde{\Gamma}(\bar{a}^{m-r}) - \tilde{\Gamma}(\tilde{a}^{m-r}) > 0$$

and this contradiction ends the proof of Theorem 4.2.

The Liapunov Function and its Properties

Let us assume that there exists an admissible detailed balance point $\hat{c}$ and consider the function $\Phi_{\hat{c}} : D \longrightarrow R$ defined by

$$\Phi_{\hat{c}}(c) = \sum_{1}^{m} c_k[\ln(c_k/\hat{c}_k) - 1], \qquad (4.2.16)$$

where $D = \{c : c_k > 0, k = 1, \ldots, m\}$ is the admissible set. The function $\Phi_{\hat{c}}$ is defined on D but it can be continuated on the closure of D (i.e. for $c_k = 0$). $\Phi_{\hat{c}}$ has the following properties:

i) It is differentiable on D;

ii) The sets $\{c : \Phi_{\hat{c}}(c) \leq \lambda\}$, where λ is a scalar, are bounded (indeed, if some $c_k \to \infty$ then, $\Phi_{\hat{c}}(c)$ is no longer bounded);

iii) The admissible detailed balance point is an absolute minimum point of $\Phi_{\hat{c}}(c)$. Indeed, from the expressions of the first order partial derivatives

$$\frac{\partial \Phi_{\hat{c}}}{\partial c_j}(c) = \ln(c_j/\hat{c}_j), \quad j = 1, \ldots, m,$$

it follows that the only critical point of $\Phi_{\hat{c}}(c)$ is $\hat{c}$ – the admissible detailed balance point whose existence has been assumed. The elements of the Hessian matrix are

$$\frac{\partial^2 \Phi_{\hat{c}}}{\partial c_i \partial c_j}(c) = \frac{1}{c_j}\delta_{ij}$$

hence this matrix is positive definite everywhere in D. The property is proved.

iv) For any $\epsilon > 0$ there exists $\delta(\epsilon)$ such that if $\Phi_{\hat{c}}(c) \leq \Phi_{\hat{c}}(\hat{c}) + \delta(\epsilon)$ then $|c - \hat{c}| < \epsilon$. Indeed

$$\Phi_{\hat{c}}(c) = \sum_{1}^{m} \hat{c}_k \Psi(c_k/\hat{c}_k)$$

where $\Psi(\lambda) = \lambda(\ln\lambda - 1)$. But $\hat{c}$ is an absolute minimum point, hence

$$\Phi_{\hat{c}}(c) - \Phi_{\hat{c}}(\hat{c}) = \sum_{1}^{m} \hat{c}_k[\Psi(c_k/\hat{c}_k) - \Psi(1)] \geq 0$$

since all terms of the sum are being nonnegative. Obviously, if $\Phi_{\hat{c}}(c) - \Phi_{\hat{c}}(\hat{c}) \leq \delta(\epsilon)$, then $\Psi(c_k/\hat{c}_k) - \Psi(1) \leq \delta_1(\epsilon)$. It remains to show that $\Psi(\lambda)$ has the following property: for any $\epsilon > 0$ there exists $\delta_1(\epsilon)$ such that if $\Psi(\lambda) - \Psi(1) \leq \delta_1(\epsilon)$ then $|\lambda - 1| < \epsilon$. Assume that this is not true, hence there exists $\epsilon > 0$ such that for any δ_1 there exists λ with $\Psi(\lambda) - \Psi(1) < \delta_1$, $|\lambda - 1| \geq \epsilon$. But $|\lambda - 1| \geq \epsilon$ means that either $\lambda \geq 1 + \epsilon$ or $\lambda \leq 1 - \epsilon$. Let $\lambda \geq 1 + \epsilon$; on this interval $\Psi(\cdot)$ is increasing:

$$\Psi(\lambda) \geq \Psi(1+\epsilon) = \Psi(1) + \frac{\epsilon^2}{2}\Psi''(\xi), \quad 1 < \xi < 1 + \epsilon$$

(recall that $\Psi'(1) = 0$, 1 being a minimum point).

From the above inequality it follows that

$$\Psi(\lambda) - \Psi(1) \geq \frac{\epsilon^2}{2} \cdot \frac{1}{\xi} > \frac{\epsilon^2}{2(1+\epsilon)}$$

which is a contradiction if $\delta_1 < \frac{\epsilon^2}{2(1+\epsilon)}$.

Let now $\lambda \leq 1 - \epsilon$; on this interval Ψ is decreasing and it follows as previously

$$\Psi(\lambda) - \Psi(1) \geq \frac{\epsilon^2}{2\xi} \geq \frac{\epsilon^2}{2}, \quad 1 - \epsilon < \xi < 1$$

– again a contradiction if $\delta_1 < \frac{\epsilon^2}{2}$.

It follows that for all δ_1 satisfying $\delta_1 < \frac{\epsilon^2}{2(1+\epsilon)}$, the inequality $|\lambda - 1| < \epsilon$ holds. Therefore the property attributed to $\Phi_{\hat{c}}$ is true.

v) Let $c(t)$ be a solution of system (4.2.2) with $c_i(0) \geq 0$ for all i. Then for any $t_2 \geq t_1 > 0$, from the existence interval of the solution, the following inequality holds:

$$\Phi_{\hat{c}}(c(t_2)) - \Phi_{\hat{c}}(c(t_1)) \leq$$

$$\leq -\sum_j \int_{t_1}^{t_2} [w_j^+(c(t)) - w_j^-(c(t))] \ln \left[\frac{w_j^+ c(t)}{w_j^- c(t)}\right] dt. \quad (4.2.17)$$

(Summation takes place only for those j for which $w_j^{\pm}(c(t)) > 0$).

The proof of this property is performed by considering the solution $c^{\epsilon}(t)$ of (4.2.2) with the initial conditions $c_i(0) + \epsilon$ for all i. It is known that $c_i^{\epsilon}(t) > 0$, hence, the mapping $t \longmapsto \Phi_{\hat{c}}(c^{\epsilon}(t))$ is well defined and smooth enough. Denoting $\tilde{\Phi}^{\epsilon}(t) = \Phi_{\hat{c}}(c^{\epsilon}(t))$ and differentiating, we find

$$\dot{\tilde{\Phi}}^{\epsilon}(t) = \sum_{i=1}^{m} \frac{\partial \Phi_{\hat{c}}}{\partial c_i}(c^{\epsilon}(t))\dot{c}_i^{\epsilon}(t) =$$

$$= \sum_{i=1}^{m} \ln(c_i^{\epsilon}(t)/\hat{c}_i) \sum_{j=1}^{n} (\beta_{ij} - \alpha_{ij}) w_j(c^{\epsilon}(t)) =$$

$$= \sum_{j=1}^{n} \left[\sum_{i=1}^{m} (\beta_{ij} - \alpha_{ij})(\ln c_i^{\epsilon}(t) - \ln \hat{c}_i)\right] w_j(c^{\epsilon}(t)).$$

But $\hat{c}$ is an admissible detailed balance point and satisfies (4.2.10). After some manipulation, it follows from the above inequality that

$$\dot{\tilde{\Phi}}^{\epsilon}(t) = -\sum_{j=1}^{n} [w_j^+(c^{\epsilon}(t)) - w_j^-(c^{\epsilon}(t))] \ln \left[\frac{w_j^+(c^{\epsilon}(t))}{w_j^-(c^{\epsilon}(t))}\right] \leq 0.$$

Therefore, for any $t_1, t_2 \in (T_1, T_2)$ it follows that

$$\tilde{\Phi}^{\epsilon}(t_2) - \tilde{\Phi}^{\epsilon}(t_1) = \Phi_{\hat{c}}(c^{\epsilon}(t_2)) - \Phi_{\hat{c}}(c^{\epsilon}(t_1)) = \quad (4.2.18)$$

$$-\sum_{j=1}^{n}\int_{t_1}^{t_2}[w_j^+(c^\epsilon(t)) - w_j^-(c^\epsilon(t))]\ln\left[\frac{w_j^+(c^\epsilon(t))}{w_j^-(c^\epsilon(t))}\right]dt \leq 0$$

Consider now a compact set included in (T_1, T_2). For any t belonging to this compact $c^\epsilon(t) \to c(t)$ when $\epsilon \to 0$, where $c_i(t) > 0$ for all i or there exist some indices for which $c_i(t) \equiv 0$ (Proposition 4.3). Consider the limit

$$\lim_{\epsilon\to 0}(w_j^+ - w_j^-)(\ln w_j^+ - \ln w_j^-)(c^\epsilon(t))$$

in the case when some $c_i(t) \equiv 0$. If $\epsilon \to 0$ then $w_j^+(c^\epsilon(t))$ tends to zero (if $\alpha_{ij} > 0$) or to some finite value (if $\alpha_{ij} = 0$); also $w_j^-(c^\epsilon(t))$ tends to zero (if $\beta_{ij} > 0$) or to some finite value (if $\beta_{ij} = 0$). Let us also note that

$$(w_j^+ - w_j^-)(\ln w_j^+ - \ln w_j^-) = (w_j^+\ln w_j^+ + w_j^-\ln w_j^-) -$$

$$-(w_j^-\ln w_j^+ + w_j^+\ln w_j^-).$$

If $w_j^+(c^\epsilon(t)) \to 0$, then $w_j^+\ln w_j^+(c^\epsilon(t)) \to 0$ and if $w_j^- \to 0$, then $w_j^-\ln w_j^- \to 0$ on the basis of the properties of function $x\ln x$. Note that $w_j^+\ln w_j^-$ has the expression

$$(w_j^+\ln w_j^-)(c^\epsilon(t)) = k_j^+\prod_{i=1}^{m}(c_i^\epsilon(t))^{\alpha_{ij}}\left[\ln k_j^- + \right.$$

$$\left. + \sum_{i=1}^{m}\beta_{ij}\ln c_i^\epsilon(t)\right]$$

and it is easy to see that if some $c_i^\epsilon \to 0$ for $\epsilon \to 0$, the limit of the above expression equals either 0 or $-\infty$ depending on the values of α_{ij} and β_{ij}. The same is true for $w_j^-\ln w_j^+$. But the RHS of (4.2) can be only finite because the LHS is finite. It follows that the limits of the expressions considered above are always zero and (4.2.17) follows.

From this property the following important consequences can be deduced:

v.1) For any solution $\Phi_{\hat{c}}(c(t)) \leq \Phi_{\hat{c}}(c(0))$ i.e. $\Phi_{\hat{c}}(c(t))$ is monotonically decreasing;

v.2) If $\Phi_{\hat{c}}(c(t_2)) = \Phi_{\hat{c}}(c(t_1))$ for some t_1, t_2 then $w_j^+(c(\tau)) \equiv w_j^-(c(\tau))$ for $t_1 \leq \tau \leq t_2$ and all j (possibly with some w_j equal to zero), that means $w_j(c(t)) \equiv 0$ for all j, hence $c(t)$ is a detailed balance point.

Remark 4.4 *If $\Phi_{\hat{c}}(c)$ and $\Gamma_{\hat{c}}(c)$ are written with respect to the same admissible point, the following equality holds*

$$\Phi_{\hat{c}}(c) + \Gamma_{\hat{c}}(c) + \sum_{k=1}^{m} \hat{c}_k \ln \hat{c}_k = \sum_{k=1}^{m} (c_k - \hat{c}_k) \ln(c_k/\hat{c}_k) \geq 0.$$

Taking into account the fact that $\Phi_{\hat{c}}(c)$ is nonincreasing along the solutions of system (4.2.2) *and that the admissible detailed balance point $\hat{c}$ is an absolute minimum point of $\Phi_{\hat{c}}$, the function*

$$V(c) = \Phi_{\hat{c}}(c) - \Phi_{\hat{c}}(\hat{c})$$

can be a Liapunov function giving at least the Liapunov stability of the equilibrium point $\hat{c}$.

Actually, the function $\Phi_{\hat{c}}$ allows one to get much more information about the qualitative behaviour of system (4.2.2); it can be considered as a Liapunov function with some degree of "universality" and attached in a natural way to the system.

Stability Results and Qualitative Properties

The main mathematical results concerning the qualitative properties of system (4.2.2) are included in

Theorem 4.3 *If system (4.2.2) is such that an admissible detailed balance point exists, the following properties of the solutions whose initial conditions satisfy $c_i(0) \geq 0$ for all i are true:*

1. *Any solution is defined and is bounded for* $T_1 < t < +\infty \quad (T_1 < 0)$
2. *There are no periodic nonconstant solutions with nonnegative components.*
3. *Any equilibrium point with nonnegative components is a detailed balance point.*
4. *The ω–limit set of any solution is composed of equilibrium points only; if such a set contains an admissible detailed balance point, it coincides with that point (i.e., it is a singleton).*
5. *An admissible detailed balance point is stable in the sense of Liapunov and it is an attractor in the invariant hyperplane which contains that point.*
6. *A solution such that* $\lim_{t\to\infty} c(t)$ *exists and has all its components positive, is stable in the sense of Liapunov.*

Proof 1) From v.1) it follows that for any solution with nonnegative initial conditions, the following inequality is true

$$\Phi_{\tilde{c}}(c(t)) \leq \Phi_{\tilde{c}}(c(0)), \quad t \geq 0$$

hence $c(t)$ is bounded. Therefore global existence follows; from Proposition 4.3 nonnegativeness of components of solution is obtained.

2) Let $c(t)$ be a periodic solution with nonnegative components. Then $c(t) = c(t+T)$ implies $\Phi_{\tilde{c}}(c(t)) = \Phi_{\tilde{c}}(c(t+T))$ which leads to $c(t) = \text{const}$.

3) Let $\tilde{c}$ be an equilibrium point of the system. From v.2) it follows that $\tilde{c}$ is a detailed balance point.

4) Let Ω_c be the ω–limit set of the solution; as it is known, this set is nonempty, compact and connected (from the boundedness of the trajectories) and invariant (composed of trajectories only). From the monotonicity of $\Phi_{\hat{c}}(c(t))$, existence of the limit $\lim_{t\to\infty} \Phi_{\hat{c}}(c(t)) = \Phi^0$ follows. Therefore, for any $\tilde{c} \in \Omega_c$ one has $\Phi_{\hat{c}}(\tilde{c}) = \Phi^0$ and, for the whole trajectory passing through $\tilde{c}$, $\Phi_{\hat{c}}(c(t, \tilde{c})) = \Phi^0$, hence $\Phi_{\hat{c}}(c(\cdot, \tilde{c}))$ is constant. It follows that Ω_c is composed of equilibrium points only. Assume now that $\tilde{c} \in \Omega_c$ is an admissible detailed balance point. It has been already proved (Theorem 4.2) that this point is unique in the linear invariant manifold (4.2.7) that contains it.

But Ω_c belongs to the manifold (4.2.7) containing the solution and if Ω_c contains the admissible detailed balance point it cannot contain other equilibrium points, being a connected set.

5) Let $\bar{c}$ be an admissible detailed balance point. Consider the function $\Phi_{\bar{c}}(c)$ defined for this point by replacing in the definition formula $\hat{c}_k$ by $\bar{c}_k$ and define then the function

$$V(c) = \Phi_{\bar{c}}(c) - \Phi_{\bar{c}}(\bar{c}).$$

Taking into account that $\bar{c}$ is a unique minimum point for $\Phi_{\bar{c}}(c)$ in D, it follows that $V(c) > 0$, $V(\bar{c}) = 0$, $c \in D$. But $V(c(t))$ is nonincreasing hence the assumptions of Liapunov Stability Theorem (Theorem 2.3 in the book) are fulfilled and the Liapunov stability of the equilibrium point follows.

For the asymptotic stability, note that all solutions are bounded hence they tend asymptotically to the ω–limit set which is composed of equilibrium points only.

Consider now the invariant hyperplane (4.2.7), where $\bar{c}$ is located; according to Theorem 4.2 this is the unique admissible detailed balance point located there. In order to obtain asymptotic stability define $\Psi_0 = \min_{\partial D} \Phi_{\bar{c}}(c)$. The sets $\{c : \Phi_{\bar{c}}(c) < \gamma < \Psi_0\}$ are bounded, invariant and contain the equilibrium point $\bar{c}$. The intersection of these sets with the invariant hyperplane containing $\bar{c}$ is also a bounded, connected and invariant set containing no other ω-limit points except $\bar{c}$. Therefore $\bar{c}$ is an attractor.

6) Let $\tilde{c}(t)$ be a solution such that $\lim_{t\to\infty} \tilde{c}(t) = \bar{c}$, $\bar{c}_k > 0$. It has been shown above that $\bar{c}$ is an admissible detailed balance point. With this equilibrium point we associate the corresponding function $\Phi_{\bar{c}}(c)$

$$\Phi_{\bar{c}}(c) = \sum_{k=1}^{m} c_k \left(\ln \frac{c_k}{\bar{c}_k} - 1 \right)$$

Property iv) of this function shows that there exists $\delta_1(\epsilon)$ such that if $\Phi_{\bar{c}}(c) \le \Phi_{\bar{c}}(\bar{c}) + \delta_1(\epsilon)$, then $|c - \bar{c}| < \frac{\epsilon}{3}$. On the other hand, $\Phi_{\bar{c}}(\tilde{c}(t))$ tends to $\Phi_{\bar{c}}(\bar{c})$ when $t \to \infty$, hence there exists $t_1(\epsilon) > 0$ such that

$$\Phi_{\bar{c}}(\tilde{c}(t)) \le \Phi_{\bar{c}}(\bar{c}) + \delta_1(\epsilon)\,, \quad t \ge t_1(\epsilon)$$

Therefore

$$|\tilde{c}(t) - \bar{c}| < \frac{\epsilon}{3}\,, \quad t \ge t_1(\epsilon)$$

Making use of Property iv) again, we find that one can choose $\delta_2(\epsilon) > \delta_1(\epsilon)$ such that if $\Phi_{\bar{c}}(c) \le \Phi_{\bar{c}}(\bar{c}) + \delta_2(\epsilon)$ then $|c - \bar{c}| < 2\frac{\epsilon}{3}$. But $\Phi_{\bar{c}}$ is uniformly continuous on compact sets. Therefore there exists $\eta_\epsilon(\cdot)$ such that if $|c_1 - c_2| < \eta_\epsilon(\alpha)$, then $|\Phi_{\bar{c}}(c_1) - \Phi_{\bar{c}}(c_2)| < \alpha$.

Using continuity with respect to initial conditions, one can find $\delta_3(\epsilon) > 0$ such that if $|c(0) - \tilde{c}(0)| < \delta_3(\epsilon)$, then

$$|c(t) - \tilde{c}(t)| < \min\{\epsilon\,, \eta_\epsilon(\delta_2(\epsilon) - \delta_1(\epsilon))\}\,, \quad 0 \le t \le t_1(\epsilon)$$

and, in particular

$$|c(t_1(\epsilon)) - \tilde{c}(t_1(\epsilon))| < \eta_\epsilon(\delta_2(\epsilon) - \delta_1(\epsilon)).$$

From the uniform continuity, it follows that

$$|\Phi_{\bar{c}}(c(t_1(\epsilon))) - \Phi_{\bar{c}}(\tilde{c}(t_1(\epsilon)))| < \delta_2(\epsilon) - \delta_1(\epsilon).$$

Therefore

$$\Phi_{\bar{c}}(c(t_1(\epsilon))) \le \Phi_{\bar{c}}(\bar{c}) + \delta_2(\epsilon)$$

But $\Phi_{\bar{c}}(c(t))$ is nonincreasing hence $\Phi_{\bar{c}}(c(t)) \leq \Phi_{\bar{c}}(\bar{c}) + \delta_2(\epsilon)$ for $t \geq t_1(\epsilon)$. It follows that $|c(t) - \bar{c}| < \frac{2\epsilon}{3}$ for $t \geq t_1(\epsilon)$ and, therefore

$$|c(t) - \tilde{c}(t)| < \epsilon, \quad t \geq t_1(\epsilon)$$

Together with the estimate for $0 \leq t \leq t_1(\epsilon)$ this gives the required property completes the proof of Theorem 4.3.

The theorem proved above shows the large amount of qualitative results that can be obtained with a suitable Liapunov function: existence and global boundedness of the solutions, information about the stationary set and the ω–limit set, periodic solutions, Liapunov stability and asymptotic stability.

It has been already mentioned that, among all possible stationary points, only the detailed balance points have a real practical importance. The theorem just proved shows that all equilibrium points with nonnegative components i.e. having physical significance are detailed balance points.

The theorem contains also some information about the ω–limit set of a physically significant solution: it consists of equilibrium points i.e. of detailed balance points only and is subject to an alternative: either equilibrium points with some zero components or a unique admissible i.e. with positive components equilibrium point. This unique admissible point is stable in the sense of Liapunov; if the solutions are restricted to the invariant hyperplane containing this point, stability is also asymptotic. Moreover, the solution tending to such an equilibrium point is Liapunov stable itself.

It is clear that a lot of properties of the admissible detailed balance points can be obtained. A problem which is open for a long time concerns attraction domain of an admissible detailed balance point in the invariant hyperplane containing it. It is not yet proved that for any solution with positive components the ω–limit set contains a point with positive components. If this conjecture were true, then the ω–limit set would not contain other equilibrium points and the attraction domain would be the whole invariant hyperplane containing the admissible equilibrium point.

It is worth mentioning as a final remark that the only assumption required for the proof of Theorem 4.3 was the existence of an admissible detailed balance point. If (4.2.12) is taken into account, it follows that this assumption introduces a restriction on the matrix of the stoichiometric coefficients and on the coefficients $k_j^{\pm}$ of reaction velocities. This condition can be easily checked in practice using a digital computer.

4.3 Processes in Plate Columns

The plate columns are chemical plants for staged countercurrent mass transfer processes. The commonly used processes of this kind are the extraction chains and the distillation processes. Both perform a contacting of two streams flowing counter to each other and the models describing the two systems behaviour are very similar in nature (J.C.Friedly, 1972; p.33). As it will become clear below the two models can be tackled together.

The basic stage of an *extraction chain* is formed by a mixer with a settler. In the mixer section the light and heavy phases have total molar holdups of h_j and H_j, assumed constant. The mole fractions of the component in question in the binary mixture are y_j' and x_j'. Complete mixing will be assumed from the outset so that all outlet concentrations equal those in the mixed fluid. Writing a component mass–balance around the entire mixer section gives

$$\frac{d}{dt}(H_j x_j' + h_j y_j') = L_{j+1}x_{j+1} - L_j x_j' + V_{j-1}y_{j-1} - V_j y_j' \tag{4.3.1}$$

where L_j and V_j are the molar flow rates of the heavy and light phases. If it is assumed that mass transfer resistance is sufficiently small, so that complete equilibrium is achieved, y_j' is related to x_j' by the *thermodynamic constitutive equation*

$$y_j' = \mu_j(x_j') \tag{4.3.2}$$

where $\mu_j : [0,1] \longrightarrow [0,1]$ are continuous and monotonically increasing. This is assumed to apply in the unsteady as well as in the steady state.

The settler portion of the stage is conveniently treated if it is assumed that no mass exchange occurs there. This is a reasonable assumption since the mixture enters in equilibrium. If it is further assumed that complete mixing occurs in each phase, the component mass balance can be written for heavy and light phases separately:

$$\frac{d}{dt}H'_j x_j = L_j(x'_j - x_j)$$

$$\frac{d}{dt}h'_j y_j = V_j(y'_j - y_j). \tag{4.3.3}$$

Equations (4.3.1) – (4.3.3) can define a complete model for the extraction chain (called sometimes extraction plate column). It should be nevertheless taken into account that there exist a heavy feed at the top of the column (the n–th stage – on the mixer) and a light feed at the bottom, of the column (the 1–st stage – on the mixer). Also there exist a light product taken from the settler at the top and a heavy product taken from the settler at the bottom. Therefore the equations for the 1–st stage are

$$\frac{d}{dt}(H_1 x'_1 + h_1 y'_1) = L_2 x_2 - L_1 x'_1 - V_1 y'_1 + F_1 z_1 \,,$$

$$\frac{d}{dt}H'_1 x_1 = L_1(x'_1 - x_1)\,,$$

$$\frac{d}{dt}h'_1 y_1 = V_1(y'_1 - y_1)\,,$$

where L_1, x_1 are here the molar flow rate and the molar fraction for the heavy product and F_1, z_1 are the same parameters for the light feed.

For the n–th stage:

$$\frac{d}{dt}(H_n x'_n + h_n y'_n) = -L_n x'_n + V_{n-1} y_{n-1} - V_n y'_n + F_n z_n \,,$$

$$\frac{d}{dt}H'_n x_n = L_n(x'_n - x_n),$$

$$\frac{d}{dt}h'_n y_n = V_n(y'_n - y_n),$$

where V_n, y_n are the molar flow rate and the molar fraction for the light product and F_n, z_n are the same parameters for the heavy feed.

Usually, an intermediate feed on the m–th stage occurs, leading to the following equation:

$$\frac{d}{dt}(H_m x'_m + h_m y'_m) = L_{m+1}x_{m+1} - L_m x'_m + V_{m-1}y_{m-1} -$$

$$-V_{m-1}y'_{m-1} + F_m z_m,$$

the other equations of the stage being unchanged.

As a rule, $h_j \ll H_j$ hence $h_j/H_j \approx 0$ and we obtain the standard equations describing the extraction chain:

$$H_j\frac{dx'_j}{dt} = L_{j+1}x_{j+1} - L_j x'_j + V_{j-1}y_{j-1} - V_j\mu_j(x'_j) + F_j z_j$$

$$H'_j\frac{dx_j}{dt} = L_j(x'_j - x_j) \tag{4.3.4}$$

$$h'_j\frac{dy_j}{dt} = V_j(\mu_j(x'_j) - y_j), \quad j = 1, 2, \ldots, n,$$

where $L_{n+1} = 0$, $V_0 = 0$ and not all $F_j = 0$ (usually $F_1 \neq 0$, $F_n \neq 0$ and sometimes $F_m \neq 0$).

In most practical situations h'_j, H'_j are negligible with respect to H_j; taking $h'_j = H'_j = 0$ one finds $x_j = x'_j$, $y_j = \mu_j(x'_j)$; the model becomes

$$H_j\frac{dx'_j}{dt} = L_{j+1}x'_{j+1} - L_j x'_j + V_{j-1}\mu_{j-1}(x'_{j-1}) -$$

$$-V_j\mu_j(x'_j) + F_j z_j, \quad j = 1, 2, \ldots, n \tag{4.3.5}$$

which is the most common in process dynamics.

Consider now the j–th stage of a distillation column which, as a rule, is considered to behave like a settler of the extraction chain. The balance equations are:

$$H_j \frac{dx_j}{dt} = L_{j+1}x_{j+1} - L_j x_j + V_{j-1}y_{j-1} - V_j y'_j$$

$$h_j \frac{dy_j}{dt} = V_j(y'_j - y_j) \qquad (4.3.6)$$

$$y'_j = \mu_j(x_j)$$

where the significance of the notations is the same as above.

The distillation column is fed by heavy, light or mixed substances on some m–th stage:

$$H_m \frac{dx_m}{dt} = L_{m+1}x_{m+1} - L_m x_m + V_{m-1}y_{m-1} - V'_m y'_m + F_m z_m\,,$$

$$h_m \frac{dy_m}{dt} = V'_m y'_m - V_m y_m + F'_m z'_m\,, \qquad (4.3.7)$$

where F_m, z_m are the parameters of the heavy feed and F'_m, z'_m are the parameters of the light feed. Sometimes the light and the heavy feed can occur on different stages.

At the top of the column, the light product is extracted but a percentage of it is cooled and recirculated as a reflux. Therefore, the equations of the n–th stage become

$$H_n \frac{dx_n}{dt} = R_n x_n - L_n x_n + V_{n-1}y_{n-1} - V_n y'_n$$

$$h_n \frac{dy_n}{dt} = V_n(y'_n - y_n)\,, \qquad (4.3.8)$$

where R_n is the reflux molar flow rate. Notice that (4.3.8) corresponds to (4.3.7) in the case $z_n = x_n$, $F'_n = 0$. The extracted light product is $V_n y_n$.

At the bottom of the column the heavy product is extracted but a percentage of it is reboiled and recirculated as a bottom reflux. Therefore, the equations of the first stage become:

$$H_1 \frac{dx_1}{dt} = L_2 x_2 - L_1 x_1 + R_1 y_1 - V_1 y_1'$$

$$h_1 \frac{dy_1}{dt} = V_1(y_1' - y_1) \tag{4.3.9}$$

where R_1 is the reflux molar flow rate.

This corresponds to (4.3.7) with $F_1 = R_1$, $z_1 = y_1$, $F_1' = 0$. The extracted heavy product is $L_1 x_1$.

In this way, the standard equations for the distillation column are obtained:

$$H_j \frac{dx_j}{dt} = L_{j+1} x_{j+1} - L_j x_j + V_{j-1} y_{j-1} - V_j' \mu_j(x_j) + F_j z_j$$

$$h_j \frac{dy_j}{dt} = V_j' \mu_j(x_j) - V_j y_j + F_j' z_j', \quad j = 1, \dots, n, \tag{4.3.10}$$

where $L_{n+1} = V_0 = 0$ and not all F_j, $F_j' = 0$ (usually $F_m \neq 0$ or $F_m' \neq 0$, $F_1 = R_1$, $z_1 = y_1$, $F_n = R_n$, $z_n = x_n$).

In most of the practical cases $h_j \ll H_j$; taking $h_j = 0$, it follows

$$H_j \frac{dx_j}{dt} = L_{j+1} x_{j+1} - L_j x_j + V_{j-1}' \mu_{j-1}(x_{j-1}) - V_j' \mu_j(x_j) +$$

$$+(F_j z_j + F_{j-1}' z_{j-1}'), \quad j = 1, \dots, n, \tag{4.3.11}$$

which is the most often encountered system describing the distillation columns dynamics (see, for instance, the books of Luyben, 1970, and Friedly, 1972). If all F_j, F_j' but F_m are zero, we have the model discussed in the paper of Gothard, Halanay and Popov (1968) (the case of the column without reflux). The distillation columns with reflux were studied from the mathematical point of view in the Ph.D.thesis of Stepan (1987) (see also papers of Reghis and Stepan, 1985 and of Balint and Stepan, 1980). We mention also that the extraction chains dynamics was studied from the mathematical point of view in the Ph.D.thesis of Burdescu (1990; see also the paper of Burdescu and Răsvan, 1990) using the model (4.3.5).

Comparing (4.3.5) and (4.3.11), we see that these two models are identical. This suggests introduction of a single model allowing to obtain all previously mentioned cases in unified way. If (4.3.4) and (4.3.10) are compared, the following general equations can be considered:

$$H_j \frac{dx'_j}{dt} = L_{j+1}x_{j+1} - L_j x'_j + V_{j-1}y_{j-1} - V'_j \mu_j(x'_j) + F_j z_j$$

$$H'_j \frac{dx_j}{dt} = L_j(x'_j - x_j) \tag{4.3.12}$$

$$h'_j \frac{dy_j}{dt} = V'_j \mu_j(x'_j) - V_j y_j + F'_j z'_j\,, \quad j = 1, 2, \ldots, n$$

with $L_{n+1} = V_0 = 0$. The other coefficients are all nonnegative due to their physical significance (L_j, F_k, F'_i, V_j, V'_j are molar flow rates, H_j, h'_j, H'_j are molar holdups and z_j, z'_j are mole fractions). The functions $\mu_j : [0, 1] \longrightarrow [0, 1]$ are continuous and increasing.

Let us note that (4.3.4) is obtained from (4.3.12) by taking $F'_j = 0$ (as it will be seen further this automatically implies $V'_j = V_j$); this corresponds to the most general case of the extraction column. The standard equations for the distillation column (4.3.10) can be obtained from (4.3.12) by taking $H'_j = 0$ hence $x_j = x'_j$. The other models can be obtained either from (4.3.4) or from (4.3.10), accordingly. This model analysis is in fact performed at the engineering level; it has only a heuristic importance because, for instance, the reduction of the small parameters multiplying the derivatives can be rigorously justified only by singular perturbations theory. For these reasons we not consider $h_j/H_j \approx 0$ any longer, and the general model of process dynamics in a plate column will have the form:

$$H_j \frac{d}{dt}\left(x'_j + \frac{h_j}{H_j}\mu_j(x'_j)\right) = L_{j+1}x_{j+1} - L_j x'_j + V_{j-1}y_{j-1} -$$

$$-V_j'\mu_j(x_j') + F_j z_j$$

$$H_j' \frac{dx_j}{dt} = L_j(x_j' - x_j) \tag{4.3.12a}$$

$$h_j' \frac{dy_j}{dt} = V_j'\mu_j(x_j') - V_j y_j + F_j' z_j' , \quad j = 1, \dots, n$$

where the significance and the sign of the coefficients are the same as in (4.3.12).

The system (4.3.12a) can be rewritten in the normal (Cauchy) form by taking into account the fact that the mapping $x \longmapsto x + \frac{h_j}{H_j}\mu_j(x)$ is invertible; a simple change of variables would solve this problem. Nevertheless, for the sake of simplicity we will assume $\mu_j(x)$ to be C^1 mappings; this allows to write (4.3.12a) as follows

$$H_j \frac{dx_j'}{dt} = \frac{1}{1 + \frac{h_j}{H_j}\frac{d\mu_j}{d\lambda}(x_j')}[L_{j+1}x_{j+1} - L_j x_j' + V_{j-1}y_{j-1} -$$

$$-V_j'\mu_j(x_j') + F_j z_j]$$

$$H_j' \frac{dx_j}{dt} = L_j(x_j' - x_j) \tag{4.3.12b}$$

$$h_j' \frac{dy_j}{dt} = V_j'\mu_j(x_j') - V_j y_j + F_j' z_j' , \quad j = 1, \dots, n$$

Two remarks are in order. The first is related to the fact that $\mu_j(x)$ are increasing, hence $\frac{d\mu_j}{d\lambda}(x) \geq 0$ and $1 + \frac{h_j}{H_j}\frac{d\mu_j}{d\lambda}(x) > 0$ for all $x \in [0,1]$. The second one points out that the assumption about $\mu_j(\cdot)$ to be C^1 mappings is not very restrictive. Indeed, in most of the practical situations $\mu_j(\cdot)$ have the form

$$\mu_j(x) = \frac{\alpha_j x}{1 - (\alpha_j - 1)x}, \quad \alpha_j > 1$$

and satisfy all the above assumptions.

Let us recall the fact that all the equations written above are obtained from component mass balances around various sections.

If global (not component) mass balances are also written around various sections, the following relations between molar flow rates i.e. between system coefficients can be obtained:

$$L_{j+1} - L_j + V_{j-1} - V'_j + F_j = 0$$

$$V'_j - V_j + F'_j = 0 \qquad n = 1, \dots, n \tag{4.3.13}$$

with $L_{n+1} = V_0 = 0$. This are design relations for the plate columns, allowing to determine some of the flow rates when other ones are given. (Usually F_j, F'_j and V_j are given and V'_j, L_j have to be computed). It is now obvious that if some $F'_j = 0$, then $V'_j = V_j$.

In what follows, we will consider the model defined by (4.3.12b) under the assumptions (4.3.13) about the coefficients.

4.3.1 The Invariant Set of the Model

The state variables of (4.3.12b) being molar fractions, have to obey the inequalities $0 \le x'_j \le 1, 0 \le x_j \le 1, 0 \le y_j \le 1$ for all j. This property is valid at any moment of time and, mathematically speaking, is a property of the system of differential equations; the above inequalities for the initial conditions should imply the fulfilment of the same inequalities along the solutions i.e. the systems should have an invariant set.

Proposition 4.4 *Consider the system* (4.3.12b) *with nonnegative coefficients* $L_{n+1} = V_0 = 0$, *the nonlinear functions* $\mu_j : [0,1] \longrightarrow [0,1]$ *being* C^1 *and monotonically increasing. Assume that the coefficients satisfy (4.3.13) and that* $0 \le z_j \le 1$, $0 \le z'_j \le 1$, $j = 1, \dots, n$. *If* $0 \le x'_j(0) \le 1$, $0 \le x_j(0) \le 1$, $0 \le y_j(0) \le 1$, *then* $0 \le x'_j(t) \le 1$, $0 \le x_j(t) \le 1$, $0 \le y_j(t) \le 1$ *for all* $t \ge 0$ *and the solution is thus defined on the whole positive semiaxis.*

In order to prove this Proposition, we recall here the following result (Pavel and Turinici, 1978).

Proposition 4.5 *Consider the continuous mapping* $F : (t_0, t_1) \times \prod_1^n [a_i, b_i] \longrightarrow R^n$. *The necessary and sufficient condition that for any* $z^0 \in \prod_1^n [a_i, b_i]$ *there exist* $T(z^0) > 0$ *and a solution* $z : [t_0, t_0 + T(z^0)] \longrightarrow R^n$ *of the Cauchy problem*

$$\dot{z} = F(t, z), \qquad z(t_0) = z^0$$

such that $z(t) \in \prod_1^n [a_i, b_i]$, $t_0 \leq t \leq t_0 + T(z^0)$, *is*

$$F_i(t, z_1, z_2, \ldots, z_{i-1}, a_i, z_{i+1}, \ldots, z_n) \geq 0$$

$$F_i(t, z_1, z_2, \ldots, z_{i-1}, b_i, z_{i+1}, \ldots, z_n) \leq 0 \quad i = 1, 2, \ldots, n$$

$$t_0 \leq t \leq t_1, \qquad a_k \leq z_k \leq b_k, \quad k \neq i$$

The proof of this Proposition is reproduced in the Appendix and is a consequence of lemma of Crandall (1973).

Proof of Proposition 4.4 We will apply the above Proposition: in our case $a_k = 0$, $b_k = 1$.

a) $$L_2 x_2 + F_1 z_1 \geq 0, \qquad 0 \leq x_2 \leq 1, \qquad 0 \leq z_1 \leq 1.$$

$$L_2 x_2 - L_1 - V_1' + F_1 z_1 = L_2 x_2 - L_2 - F_1 + F_1 z_1 =$$

$$= -L_2(1 - x_2) - F_1(1 - z_1) \leq 0, \qquad 0 \leq x_2 \leq 1,$$

$$0 \leq z_1 \leq 1.$$

$$L_{j+1} x_{j+1} + V_{j-1} y_{j-1} + F_j z_j \geq 0, \qquad 0 \leq x_{j+1} \leq 1,$$

$$0 \leq y_{j-1} \leq 1, \qquad 0 \leq z_j \leq 1$$

$$L_{j+1} x_{j+1} - L_j - V_j' + V_{j-1} y_{j-1} + F_j z_j =$$

$$= L_{j+1} x_{j+1} - L_{j+1} - V_{j-1} - F_j + V_{j-1} y_{j-1} + F_j z_j =$$

$$= -L_{j+1}(1 - x_{j+1}) - V_{j-1}(1 - y_{j-1}) - F_j(1 - z_j) \leq 0,$$

$$0 \leq x_{j+1} \leq 1, \quad 0 \leq y_{j-1} \leq 1, \quad 0 \leq z_j \leq 1,$$

$$j = 2, \ldots, n-1$$

$$V_{n-1}y_{n-1} + F_n z_n \geq 0, \quad 0 \leq y_{n-1} \leq 1, \quad 0 \leq z_n \leq 1$$

$$-L_n - V'_n + V_{n-1}y_{n-1} + F_n z_n =$$

$$= -V_{n-1}(1 - y_{n-1}) - F_n(1 - z_n) \leq 0,$$

$$0 \leq y_{n-1} \leq 1, \qquad 0 \leq z_n \leq 1.$$

b) $$L_j(x'_j - 0) \geq 0, \quad 0 \leq x'_j \leq 1, \quad j = 1, \ldots, n$$

$$L_j(x'_j - 1) \leq 0, \quad 0 \leq x'_j \leq 1, \quad j = 1, \ldots, n$$

c) $$V'_j \mu_j(x'_j) + F'_j z'_j \geq 0, \quad 0 \leq x'_j \leq 1, \quad 0 \leq z'_j \leq 1,$$

$$j = 1, \ldots, n$$

$$V'_j \mu_j(x'_j) - V_j + F'_j z'_j = -V'_j(1 - \mu_j(x'_j)) - F'_j(1 - z'_j) \leq 0,$$

$$0 \leq x'_j \leq 1, \quad 0 \leq z'_j \leq 1, \quad j = 1, \ldots, n$$

We have thus checked the fulfilment of the conditions of the above Proposition by system (4.3.12b); use was made essentially of (4.3.13). Suince all the conditions of the Proposition are fulfilled for (4.3.12b) the main result of Proposition 4.4 follows. The fact that the solutions of (4.3.12b) remain in the compact set $[0, 1]^{3n}$ ensures global existence of these solutions. Proposition 4.4 is thus completely proved.

Steady State Solutions

From the engineering point of view the steady state solutions of the system (4.3.12b) represent the so–called operating points of the plant which are essential for efficient exploitation. From the mathematical point of view, the steady state solutions are the constant solutions of (4.3.12b). Consequently, they are the solutions of the following system

$$
\begin{aligned}
&L_{j+1}x_{j+1} - L_j x'_j + V_{j-1}y_{j-1} - V'_j\mu_j(x'_j) + F_j z_j = 0\\
&x'_j - x_j = 0 \qquad\qquad (4.3.14)\\
&V'_j\mu_j(x'_j) - V_j y_j + F'_j z'_j = 0\,, \quad j = 1,\dots,n.
\end{aligned}
$$

In fact, not all solutions of (4.3.14) are of interest but only those belonging to the hypercube $[0,1]^{3n}$; this is due again to the fact that x'_j, x_j, y_j are mole fractions and the mathematical steady state solution should correspond to its physical meaning.

The last two equations of (4.3.14) are linear with respect to the unknown variables x_j and y_j, hence for given z'_j, x'_j they are determined in unique way:

$$
\begin{aligned}
&x^*_j = x'_j\\
&y^*_j = \frac{1}{V_j}(V'_j\mu_j(x'_j) + F'_j z'_j) \qquad\qquad (4.3.15)
\end{aligned}
$$

Assuming that $0 \le x'_j \le 1$, it follows that $0 \le x^*_j \le 1$ and taking into account that μ_j are monotonically increasing,

$$y^*_j \ge \frac{1}{V_j}(V'_j\mu_j(0) + F'_j z'_j) \ge \frac{F'_j}{V_j} z'_j \ge 0$$

and

$$y^*_j \le \frac{1}{V_j}(V'_j\mu_j(1) + F'_j z'_j) \le \frac{1}{V_j}(V'_j + F'_j z'_j) = 1 - \frac{F'_j}{V_j}(1 - z'_j) \le 1.$$

In order to establish the intervals where x_j^{**}, the steady–state values for x'_j, are located, we substitute $x_j^*(x'_j)$ and $y_j^*(x'_j)$ in the first equation of (4.3.14). Therefore the following nonlinear system is obtained:

$$L_2x'_2 - L_1x'_1 - V'_1y'_1 + F_1z_1 = 0$$

$$L_3x'_3 - L_2x'_2 + V'_1y'_1 - V'_2y'_2 + F'_1z'_1 + F_2z_2 = 0$$

$$- -$$

$$L_{j+1}x'_{j+1} - L_jx'_j + V'_{j-1}y'_{j-1} - V'_jy'_j + F'_{j-1}z'_{j-1} + F_jz_j = 0$$

$$- -$$

$$L_nx'_n - L_{n-1}x'_{n-1} + V'_{n-2}y'_{n-2} - V'_{n-1}y'_{n-1} +$$

$$+F'_{n-2}z'_{n-2} + F_{n-1}z_{n-1} = 0$$

$$-L_nx'_n + V'_{n-1}y'_{n-1} - V'_ny'_n + F'_{n-1}z'_{n-1} + F_nz_n = 0$$

where

$$y'_j = \mu_j(x'_j), \qquad j = 1, \ldots, n.$$

It follows from the last equality of the steady–state system

$$y'_{n-1} = \frac{1}{V'_{n-1}}[L_nx'_n + V'_n\mu_n(x'_n) - F_nz_n - F'_{n-1}z'_{n-1}]$$

Introducing the function

$$\eta_{n-1}(u, v) = \frac{1}{V'_{n-1}}[L_nu + V'_n\mu_n(v) - F_nz_n - F'_{n-1}z'_{n-1}]$$

we have

$$y'_{n-1} = \eta_{n-1}(x'_n, x'_n) = \gamma_{n-1}(x'_n)$$

where $\gamma_{n-1}(\cdot)$ is a monotonically increasing mapping of the interval [0,1] on $\left[-\frac{F_n z_n + F'_{n-1} z'_{n-1}}{V'_{n-1}}, 1 + \frac{F_n(1-z_n)+F'_{n-1}(1-z'_{n-1})}{V'_{n-1}}\right]$ which includes the interval [0,1]. Defining the numbers $\underline{x}^n$ and $\bar{x}^n$ from the equalities

$$\gamma_{n-1}(\underline{x}^n) = 0, \qquad \gamma_{n-1}(\bar{x}^n) = 1,$$

the mapping $\gamma_{n-1}(\cdot)$ maps the interval $[\underline{x}^n, \bar{x}^n]$ on [0,1]. Here obviously $0 < \underline{x}^n < \bar{x}^n < 1$. Therefore, if $x_n^{**} \in [\underline{x}^n, \bar{x}^n]$ then $y_{n-1}^{**} = \gamma_{n-1}(x_n^{**}) \in [0,1]$ and $x_{n-1}^{**} = \mu_{n-1}^{-1}(y_{n-1}^{**}) = (\mu_{n-1}^{-1} \circ \gamma_{n-1})(x_n^{**}) = \varphi_{n-1}(x_n^{**}) \in [0,1]$.

Adding the last two equations, it follows

$$-L_{n-1}x'_{n-1} + V'_{n-2}y'_{n-2} - V'_n y'_n + F_n z_n + F_{n-1}z_{n-1} +$$

$$+F'_{n-1}z'_{n-1} + F'_{n-2}z'_{n-2} = 0$$

hence

$$y'_{n-2} = \frac{1}{V'_{n-2}}\left[L_{n-1}x'_{n-1} + V'_n y'_n - \sum_{j=n-1}^{n} F_j z_j - \sum_{j=n-2}^{n-1} F'_j z'_j\right]$$

Introducing the function

$$\eta_{n-2}(u,v) = \frac{1}{V'_{n-2}}\left[L_{n-1}u + V'_n \mu_n(v) - \sum_{k=n-1}^{n} F_k z_k - \sum_{j=n-2}^{n-1} F'_j z'_j\right]$$

we have

$$y'_{n-2} = \eta_{n-2}(\varphi_{n-1}(x'_n), x'_n) \equiv \gamma_{n-2}(x'_n)$$

where $\gamma_{n-2}(\cdot)$ is a monotonically increasing mapping of the interval $[\underline{x}^n, \bar{x}^n]$ on some interval that contains the interval [0,1]. Defining the numbers $\underline{x}^{n-1}$, $\bar{x}^{n-1}$ from the equalities

$$\gamma_{n-2}(\underline{x}^{n-1}) = 0, \qquad \gamma_{n-2}(\bar{x}^{n-1}) = 1,$$

the mapping $\gamma_{n-2}(\cdot)$ maps the interval $[\underline{x}^{n-1}, \bar{x}^{n-1}]$ on [0,1]. Obviously $0 < \underline{x}^n < \underline{x}^{n-1} < \bar{x}^{n-1} < \bar{x}^n < 1$.

Therefore, if $x_n^{**} \in [\underline{x}^{n-1}, \bar{x}^{n-1}]$ then

$$y_{n-1}^{**} = \gamma_{n-1}(x_n^{**}) \in [\gamma_{n-1}(\underline{x}^{n-1}), \gamma_{n-1}(\bar{x}^{n-1})] \subset [0,1],$$

$$x_{n-1}^{**} = \varphi_{n-1}(x_n^{**}) \in [\varphi_{n-1}(\underline{x}^{n-1}), \varphi_{n-1}(\bar{x}^{n-1})] \subset [0,1],$$

$$y_{n-2}^{**} = \gamma_{n-2}(x_n^{**}) \in [0,1],$$

$$x_{n-2}^{**} = \varphi_{n-2}(x_n^{**}) \in [0,1].$$

Here $\varphi_{n-2}(x) = (\mu_{n-2}^{-1} \circ \gamma_{n-2})(x)$

In the general case, adding the last k equations, we find

$$-L_{n-k+1}x'_{n-k+1} - V'_n y'_n + V'_{n-k} y'_{n-k} + \\ + \sum_{j=n-k+1}^{n} F_j z_j + \sum_{j=n-k}^{n-1} F'_j z'_j = 0,$$

hence

$$y'_{n-k} = \frac{1}{V'_{n-k}} \left[L_{n-k+1} x'_{n-k+1} + V'_n \mu_n(x'_n) - \sum_{j=n-k+1}^{n} F_j z_j - \right. \\ \left. - \sum_{j=n-k}^{n-1} F'_j z'_j \right]$$

and we can define

$$\eta_{n-k}(u,v) = \frac{1}{V'_{n-k}} \left[L_{n-k+1} u + V'_n \mu_n(v) - \sum_{j=n-k+1}^{n} F_j z_j - \right. \\ \left. - \sum_{j=n-k}^{n-1} F'_j z'_j \right]. \tag{4.3.16}$$

Therefore

$$y'_{n-k} = \eta_{n-k}(\varphi_{n-k+1}(x'_n), x'_n) = \gamma_{n-k}(x'_n) \tag{4.3.17}$$

where $\gamma_{n-k}(\cdot)$ defined above is a monotonically increasing mapping of the interval $[\underline{x}^{n-k+2}, \bar{x}^{n-k+2}]$ on some interval which contains the interval [0,1]. Defining the numbers $\underline{x}^{n-k+1}$, $\bar{x}^{n-k+1}$ from the equalities

$$\gamma_{n-k}(\underline{x}^{n-k+1}) = 0, \qquad \gamma_{n-k}(\bar{x}^{n-k+1}) = 1$$

the mapping $\gamma_{n-k}(\cdot)$ maps the interval $[\underline{x}^{n-k+1}, \bar{x}^{n-k+1}]$ on $[0, 1]$. Obviously

$$0 < \underline{x}^{n} < \underline{x}^{n-1} < \ldots < \underline{x}^{n-k+1} < \bar{x}^{n-k+1} <$$

$$< \bar{x}^{n-k+2} < \ldots < \bar{x}^{n-1} < \bar{x}^{n} < 1$$

Therefore, if $x_n^{**} \in [\underline{x}^{n-k+1}, \bar{x}^{n-k+1}]$ then

$$y_{n-1}^{**} = \gamma_{n-1}(x_n^{**}) \in [\gamma_{n-1}(\underline{x}^{n-k+1}), \gamma_{n-1}(\bar{x}^{n-k+1})] \subset [0, 1]$$

$$x_{n-1}^{**} = \varphi_{n-1}(x_n^{**}) \in [\varphi_{n-1}(\underline{x}^{n-k+1}), \varphi_{n-1}(\bar{x}^{n-k+1})] \subset [0, 1]$$

$$- \qquad (4.3.18)$$

$$y_{n-k+1}^{**} = \gamma_{n-k+1}(x_n^{**}) \in$$

$$\in [\gamma_{n-k+1}(\underline{x}^{n-k+1}), \gamma_{n-k+1}(\bar{x}^{n-k+1})] \subset [0, 1]$$

$$x_{n-k+1}^{**} = \varphi_{n-k+1}(x_n^{**}) \in$$

$$\in [\varphi_{n-k+1}(\underline{x}^{n-k+1}), \varphi_{n-k+1}(\bar{x}^{n-k+1})] \subset [0, 1]$$

$$y_{n-k}^{**} = \gamma_{n-k}(x_n^{**}) \in [\gamma_{n-k}(\underline{x}^{n-k+1}), \gamma_{n-k}(\bar{x}^{n-k+1})] \subset [0, 1]$$

$$x_{n-k}^{**} = \varphi_{n-k}(x_n^{**}) \in [\varphi_{n-k}(\underline{x}^{n-k+1}), \varphi_{n-k}(\bar{x}^{n-k+1})] \subset [0, 1]$$

Here $\varphi_{n-k}(x) = (\mu_{n-k}^{-1} \circ \gamma_{n-k})(x)$.

Consider now the first equality of the steady state (the first plate). We have

$$x_2' = \frac{1}{L_2}[L_1 x_1' + V_1' \mu_1(x_1') - F_1 z_1]$$

and, introducing the function

$$\zeta_2(u,v) = \frac{1}{L_2}[L_1 u + V_1' \mu_1(v) - F_1 z_1],$$

it follows that

$$x_2' = \zeta_2(x_1', x_1') = \psi_2(x_1')$$

where $\psi_2(\cdot)$ is monotonically increasing mapping of the interval [0,1] on the interval $\left[-\frac{F_1 z_1}{L_2}, 1 + \frac{F_1}{L_2}(1 - z_1)\right]$ which includes [0,1].

Defining the numbers $\underline{x}^1$ and $\bar{x}^1$ from the equalities

$$\psi_2(\underline{x}^1) = 0, \qquad \psi_2(\bar{x}^1) = 0$$

obviously $0 < \underline{x}^1 < \bar{x}^1 < 1$ and $\psi_2(\cdot)$ maps $[\underline{x}^1, \bar{x}^1]$ on $[0, 1]$. Therefore, if $x_1^{**} \in [\underline{x}^1, \bar{x}^1]$ then $x_2^{**} = \psi_2(x_1^{**}) \in [0, 1]$.

Adding the first two equalities of the steady state, we find

$$-L_1 x_1' + L_3 x_3' - V_2' y_2' + F_1 z_1 + F_2 z_2 + F_1' z_1' = 0$$

Therefore

$$x_3' = \frac{1}{L_3}[L_1 x_1' + V_2' \mu_2(x_2') - F_1 z_1 - F_2 z_2 - F_1' z_1']$$

and, denoting

$$\zeta_3(u,v) = \frac{1}{L_3}\left[L_1 u + V_2' \mu_2(v) - \sum_{j=1}^{2} F_j z_j - F_1' z_1'\right],$$

we have

$$x_3' = \zeta_3(x_1', \psi_2(x_1')) = \psi_3(x_1')$$

where $\psi_3(\cdot)$ is a monotonically increasing mapping of the interval $[\underline{x}^1, \bar{x}^1]$ on some interval which contains the interval [0,1]. Defining the numbers $\underline{x}^2$, $\bar{x}^2$ from the equalities

$$\psi_3(\underline{x}^2) = 0, \qquad \psi_3(\bar{x}^2) = 1,$$

the mapping $\psi_3(\cdot)$ maps the interval $[\underline{x}^2, \bar{x}^2]$ on [0,1]; obviously $0 < \underline{x}^1 < \underline{x}^2 < \bar{x}^2 < \bar{x}^1 < 1$. Therefore, if $x_1^{**} \in [\underline{x}^2, \bar{x}^2]$ then

$$x_2^{**} = \psi_2(x_1^{**}) \in [\psi_2(\underline{x}^2), \psi_2(\bar{x}^2)] \subset [0,1],$$

$$x_3^{**} = \psi_3(x_1^{**}) \in [\psi_3(\underline{x}^2), \psi_3(\bar{x}^2)] = [0,1].$$

In general, adding the first $j-1$ equalities, it follows

$$-L_1 x_1' + L_j x_j' - V_{j-1}' y_{j-1}' + \sum_{k=1}^{j-1} F_k z_k + \sum_{k=1}^{j-2} F_k' z_k' = 0$$

hence

$$x_j' = \frac{1}{L_j}\left[L_1 x_1' + V_{j-1}' \mu_{j-1}(x_{j-1}') - \sum_{k=1}^{j-1} F_k z_k - \sum_{k=1}^{j-2} F_k' z_k'\right].$$

Introducing the function

$$\zeta_j(u,v) = \frac{1}{L_j}\left[L_1 u + V_{j-1}' \mu_{j-1}(v) - \sum_{k=1}^{j-1} F_k z_k - \sum_{k=1}^{j-2} F_k' z_k'\right] \quad (4.3.19)$$

we have

$$x_j' = \zeta_j(x_1', \psi_{j-1}(x_1')) = \psi_j(x_1') \quad (4.3.20)$$

where $\psi_j(\cdot)$ is a monotonically increasing mapping of the interval $[\underline{x}^{j-2}, \bar{x}^{j-2}]$ on some interval which contains the interval [0,1]. Defining the numbers $\underline{x}^{j-1}$, $\bar{x}^{j-1}$ from the equalities

$$\psi_j(\underline{x}^{j-1}) = 0, \qquad \psi_j(\bar{x}^{j-1}) = 1.$$

the mapping $\psi_j(\cdot)$ maps $[\underline{x}^{j-1}, \bar{x}^{j-1}]$ on [0,1]. Obviously

$$0 < \underline{x}^1 < \underline{x}^2 < \ldots < \underline{x}^{j-1} < \bar{x}^{j-1} < \bar{x}^{j-2} < \ldots < \bar{x}^2 < \bar{x}^1 < 1.$$

Therefore, if $x_1^{**} \in [\underline{x}^{j-1}, \bar{x}^{j-1}]$, then

$$x_2^{**} = \psi_2(x_1^{**}) \in [\psi_2(\underline{x}^{j-1}), \psi_2(\bar{x}^{j-1})] \subset [0,1]$$

$$x_3^{**} = \psi_3(x_1^{**}) \in [\psi_3(\underline{x}^{j-1}), \psi_3(\bar{x}^{j-1})] \subset [0,1]$$

$$-------------------$$

$$x_j^{**} = \psi_j(x_1^{**}) \in [\psi_j(\underline{x}^{j-1}), \psi_j(\bar{x}^{j-1})] = [0,1].$$

Consider now the m–th plate, having some significance (the intermediate feed stage for the extraction chain, the feed stage for the distillation column). Applying the above procedures we have

A – the numbers $\underline{x}^{m+1}$, $\bar{x}^{m+1}$ defined by

$$\gamma_m(\underline{x}^{m+1}) = 0\,, \qquad \gamma_m(\bar{x}^{m+1}) = 1$$

are such that

$$0 < \underline{x}^n < \ldots < \underline{x}^{m+1} < \bar{x}^{m+1} < \ldots < \bar{x}^n < 1$$

and

$$x_{n-1}^{**} = \varphi_{n-1}(x_n^{**}) \in [\varphi_{n-1}(\underline{x}^{m+1}), \varphi_{n-1}(\bar{x}^{m+1})] \subset [0,1]$$

$$--------------------- \qquad (4.3.21)$$

$$x_m^{**} = \varphi_m(x_n^{**}) \in [\varphi_m(\underline{x}^{m+1}), \varphi_m(\bar{x}^{m+1})] = [0,1]$$

where γ_k and φ_k are those defined above;

B – the numbers $\underline{x}^{m-1}$, $\bar{x}^{m-1}$ defined by

$$\psi_m(\underline{x}^{m-1}) = 0\,, \qquad \psi_m(\bar{x}^{m-1}) = 1$$

are such that

$$0 < \underline{x}^1 < \underline{x}^2 < \ldots < \underline{x}^{m-1} < \bar{x}^{m-1} < \ldots < \bar{x}^2 < \bar{x}^1 < 1$$

and

$$x_2^{**} = \psi_2(x_1^{**}) \in [\psi_2(\underline{x}^{m-1}), \psi_2(\bar{x}^{m-1})] \subset [0,1]$$

$$------------------- \qquad (4.3.22)$$

$$x_m^{**} = \psi_m(x_1^{**}) \in [\psi_m(\underline{x}^{m-1}), \psi_m(\bar{x}^{m-1})] = [0,1]$$

where ψ_k are those defined above.

In this way, all steady state values x_k^{**} for the state variables x'_k have been determined as function either of x_n^{**} (the steady state value for x'_n) or of x_1^{**} (the steady state value for x'_1); two expressions for x_m^{**} have been obtained

$$x_m^{**} = \varphi_m(x_n^{**}) = \psi_m(x_1^{**})$$

with $\varphi_m : [\underline{x}^{m+1}, \bar{x}^{m+1}] \longrightarrow [0,1]$ and $\psi_m : [\underline{x}^{m-1}, \bar{x}^{m-1}] \longrightarrow [0,1]$, φ_m and ψ_m being onto mappings. Therefore

$$x_1^{**} = (\psi_m^{-1} \circ \varphi_m)(x_n^{**}) \tag{4.3.23}$$

where $\psi_m^{-1} \circ \varphi_m$ maps $[\underline{x}^{m+1}, \bar{x}^{m+1}]$ on $[\underline{x}^{m-1}, \bar{x}^{m-1}]$.
Therefore

$$\underline{x}^{m-1} = (\psi_m^{-1} \circ \varphi_m)(\underline{x}^{m+1}), \quad \bar{x}^{m-1} = (\psi_m^{-1} \circ \varphi_m)(\bar{x}^{m+1}).$$

We add now all steady state equations

$$-L_1 x'_1 - V'_n y'_n + \sum_{k=1}^{n} F_k z_k + \sum_{k=1}^{n-1} F'_k z'_k = 0.$$

This equation can be used to obtain explicitely x_n^{**}, the steady state value of the state variable x'_n.
Indeed, we have

$$L_1 \psi_m^{-1}(\varphi_m(x_n^{**})) + V'_n \mu_n(x_n^{**}) = \sum_{k=1}^{n} F_k z_k + \sum_{k=1}^{n-1} F'_k z'_k$$

The function

$$\Phi(x) = L_1(\psi_m^{-1} \circ \varphi_m)(x) + V'_n \mu_n(x) \tag{4.3.24}$$

is monotonically increasing and x_n^{**} could be the unique solution of the equation

$$\Phi(x) = \sum_{k=1}^{n} F_k z_k + \sum_{k=1}^{n-1} F'_k z'_k. \tag{4.3.25}$$

This solution, if exists, should be located in the interval $[\underline{x}^{m+1}, \bar{x}^{m+1}]$ in order guarantee that x_k^{**} $(k = 1, \ldots, n-1)$ is located in the hypercube $[0,1]^n$.

Note that

$$\Phi(\underline{x}^{m+1}) - \sum_{k=1}^{n} F_k z_k - \sum_{k=1}^{n-1} F'_k z'_k =$$

$$= L_1 \psi_m^{-1}(\varphi_m(\underline{x}^{m+1})) + V'_n \mu_n(\underline{x}^{m+1}) - \sum_{k=1}^{n} F_k z_k - \sum_{k=1}^{n-1} F'_k z'_k =$$

$$= L_1 \underline{x}^{m-1} + V'_n \mu_n(\underline{x}^{m+1}) - \sum_{k=1}^{n} F_k z_k - \sum_{k=1}^{n-1} F'_k z'_k.$$

Using the definition of $\underline{x}^{m-1}$ and $\underline{x}^{m+1}$ it follows that

$$\Phi(\underline{x}^{m+1}) - \sum_{k=1}^{n} F_k z_k - \sum_{k=1}^{n-1} F'_k z'_k = -V'_{m-1} \mu_{m-1}(\psi_{m-1}(\underline{x}^{m-1})) -$$

$$- L_{m+1} \varphi_{m+1}(\underline{x}^{m+1}) - F_m z_m - F'_{m-1} z'_{m-1} < 0.$$

We have also

$$\Phi(\bar{x}^{m+1}) - \sum_{k=1}^{n} F_k z_k - \sum_{k=1}^{n-1} F'_k z'_k = L_1 \bar{x}^{m-1} + V'_n \mu_n(\bar{x}^{m+1}) -$$

$$- \sum_{k=1}^{n} F_k z_k - \sum_{k=1}^{n-1} F'_k z'_k$$

and, using the definitions of $\bar{x}^{m-1}$ and $\bar{x}^{m+1}$ it follows that

$$\Phi(\bar{x}^{m+1}) - \sum_{k=1}^{n} F_k z_k - \sum_{k=1}^{n-1} F'_k z'_k = L_{m+1}(1 - \varphi_{m+1}(\bar{x}^{m+1})) +$$

$$+ V'_{m-1}(1 - \mu_{m-1}(\psi_{m-1}(\bar{x}^{m-1}))) + F_m(1 - z_m) + F'_{m-1}(1 -$$

$$- z'_{m-1}) > 0.$$

The two inequalities obtained above show that (4.3.25) has indeed a unique solution located in the interval $[\underline{x}^{m+1}, \bar{x}^{m+1}]$, therefore the steady state values x_k^{**} of the state variables x'_k belong to $[0,1]^n$.

We have in fact proved.

Theorem 4.4 *Consider the system of nonlinear equations* (4.3.14) *with nonnegative coefficients satisfying* (4.3.13) *with* $L_{n+1} = V_0 = 0$. *The nonlinear functions* $\mu_j(\cdot)$ *are monotonically increasing mappings of* $[0, 1]$ *onto* $[0, 1]$. *Then the considered system has a unique solution in the hypercube* $[0, 1]^{3n}$ *for any* $0 \leq z_i \leq 1$, $0 \leq z_i' \leq 1$, $i = 1, \ldots, n$.

The importance of this result cannot be underestimated. Indeed, it has been proved that all solutions of (4.3.12b) having physical significance belong to the hypercube $[0, 1]^{3n}$. (More precisely, if they start inside this hypercube they remain there). This property does not imply the same thing for the steady state. *If the steady state* were not located in the hypercube, the system would not have physically significant steady state and this would have meant uselessness of the mathematical model because it is known that plate columns do have steady states i.e. operating points.

Stability of Steady State

Since the significance of the steady state stability is clear now, we formulate the main stability result.

Theorem 4.5 *Consider the system* (4.3.12b) *describing the dynamics of a plate column. Assume that* (4.3.13) *hold, that* $L_{n+1} = V_0 = 0$, $0 \leq z_i \leq 1$, $0 \leq z_i' \leq 1$, $i = 1, \ldots, n$, *and that the monotonically increasing* C^1 *mappings* $\mu_i(\cdot)$ *map* $[0, 1]$ *onto* $[0, 1]$. *Then this system has a unique steady state* (x_k^*, x_k^*, y_k^*), $k = 1, \ldots, n$ *belonging to the hypercube* $[0, 1]^{3n}$ *and this steady state is globally exponentially stable (here globally means "for any initial condition in the hypercube* $[0, 1]^{3n}$*").*

Before proving this result some comments are necessary. First, the existence of the unique steady state is ensured by Theorem 4.4 just proved. Because of the first equality of (4.3.15) we have $x_k^{**} = x_k^*$ and from here the structure (x_k^*, x_k^*, y_k^*) of the steady state.

Second, it has to be mentioned that for proving stability three approaches are possible: the approach of Gothard, Halanay and Popov (1968) who found out a family of intervals which converge to the steady state as well as a quadratic Liapunov function, the approach of Reghis and Stepan (1985) based on the stability by the first approximation and, finally, the approach of Rosenbrock who for a quite different model used a non–smooth Liapunov function. Here we will take the approach of constructing a non–smooth Rosenbrock–type Liapunov function for our specific model.

Proof of Theorem 4.5. As it has been already pointed out, the existence of a unique steady state within the hypercube $[0,1]^{3n}$ follows from Theorem 4.4 just proved; this is obviously due to the fact that the assumptions of Theorem 4.4 are included among the assumptions of the theorem we are now proving.

In order to prove stability of the steady state we consider the following Liapunov function

$$\mathcal{V}(x',x,y) = \sum_{i=1}^{n} |L_{i+1}x_{i+1} - L_i x_i' - V_i'\mu_i(x_i') + V_{i-1}y_{i-1} +$$

$$+F_i z_i| + \sum_{j=1}^{n} L_j |x_j' - x_j| + \sum_{k=1}^{n} |V_k'\mu_k(x_k') - V_k y_k + F_k' z_k'| \qquad (4.3.26)$$

where $x' = \mathrm{col}(x_1', \ldots, x_n')$, $x = \mathrm{col}(x_1, \ldots, x_n)$, $y = \mathrm{col}(y_1, \ldots, y_n)$.

Taking into account the equations (4.3.14) of the stationary point, we find that:

$$\mathcal{V}(x',x,y) = \sum_{i=1}^{n} |L_{i+1}(x_{i+1} - x_{i+1}^*) - L_i(x_i' - x_i^*) -$$

$$-V_i'(\mu_i(x_i') - \mu_i(x_i^*)) + V_{i-1}(y_{i-1} - y_{i-1}^*)| +$$

$$+\sum_{j=1}^{n} L_j |(x_j' - x_j^*) - (x_j - x_j^*)| + \qquad (4.3.26a)$$

$$\sum_{k=1}^{n} |V_k'(\mu_k(x_k') - \mu_k(x_k^*)) - V_k(y_k - y_k^*)|$$

hence $\mathcal{V}(x^*, x^*, y^*) = 0$ and this is the only point where the Liapunov function vanishes. Therefore, $\mathcal{V}(X) > 0$ for $X \neq X^*$ where $X = \mathrm{col}(x', x, y)$. We deduce the existence of two monotonically increasing functions α, β, $\alpha(0) = \beta(0) = 0$ such that

$$\alpha(|X - X^*|) \leq \mathcal{V}(X) \leq \beta(|X - X^*|).$$

It is also obvious that $\mathcal{V}(X)$ is Lipschitz within the hypercube $[0,1]^{3n}$. Defining $\tilde{\mathcal{V}}(t) = \mathcal{V}(X(t))$ and taking into account the structure of $\mathcal{V}$ as defined by (4.3.26a), we find that

$$D^+\tilde{\mathcal{V}}(t) = \limsup_{h\to 0+} \frac{\tilde{\mathcal{V}}(t+h) - \tilde{\mathcal{V}}(t)}{h} = \lim_{h\to 0+} \frac{\tilde{\mathcal{V}}(t+h) - \tilde{\mathcal{V}}(t)}{h}$$

because in this case the limit does exist.

We know also that

$$\lim_{h\to 0+} \frac{|f(t+h)| - |f(t)|}{h} = \sigma(t) f'(t)$$

where

$$\sigma(t) = \begin{cases} +1 & \text{if } f(t) > 0 \text{ or } f(t) = 0 \text{ and } f'(t) > 0 \\ 0 & \text{if } f(t) = 0 \text{ and } f'(t) = 0 \\ -1 & \text{if } f(t) < 0 \text{ or } f(t) = 0 \text{ and } f'(t) < 0 \end{cases} \qquad (4.3.27)$$

Therefore

$$D^+\tilde{\mathcal{V}}(t) = \sum_{i=1}^{n} \sigma_i(t) \left\{ \frac{-(L_i + V_i' \frac{d\mu_i}{d\lambda}(x_i'(t)))}{H_i + h_i \frac{d\mu_i}{d\lambda}(x_i'(t))} [L_{i+1} x_{i+1}(t) - \right.$$

$$-L_i x_i'(t) - V_i' \mu_i(x_i'(t)) + V_{i-1} y_{i-1}(t) + F_i z_i] + \frac{L_{i+1}^2}{H_{i+1}'} [x_{i+1}'(t) -$$

$$\left. - x_{i+1}(t)] + \frac{V_{i-1}}{h_{i-1}'} [V_{i-1}' \mu_{i-1}(x_{i-1}'(t)) - V_{i-1} y_{i-1}(t) + F_{i-1}' z_{i-1}'] \right\} +$$

$$+ \sum_{j=1}^{n} \tilde{\sigma}_j(t) \left\{ \frac{L_j}{H_j + h_j \frac{d\mu_j}{d\lambda}(x_j'(t))} [L_{j+1} x_{j+1}(t) - L_j x_j'(t) - \right.$$

$$- V'_j\mu_j(x'_j(t)) + V_{j-1}y_{j-1}(t) + F_j z_j] - \frac{L_j^2}{H'_j}[x'_j(t) - x_j(t)]\Big\} +$$

$$+\sum_{k=1}^{n} \hat{\sigma}_k(t)\Big\{\frac{V'_k\frac{d\mu_k}{d\lambda}(x'_k(t))}{H_k + h_k\frac{d\mu_k}{d\lambda}(x'_k(t))}[L_{k+1}x_{k+1}(t) - L_k x'_k(t) -$$

$$-V'_k\mu_k(x'_k(t)) + V_{k-1}y_{k-1}(t) + F_k z_k] -$$

$$- \frac{V_k}{h'_k}[V'_k\mu_k(x'_k(t)) - V_k y_k(t) + F'_k z'_k]\Big\},$$

where $\sigma_i(t)$, $\tilde{\sigma}_j(t)$, $\hat{\sigma}_k(t)$, $i, j, k = 1, 2, \ldots, n$, are defined by (4.3.27), being associated to $L_{i+1}x_{i+1}(t) - L_i x_i(t) - V'_i\mu_i(x'_i(t)) + V_{i-1}y_{i-1}(t) + F_i z_i$; $L_j[x'_j(t) - x_j(t)]$, $V'_k\mu_k(x'_k(t)) - V_k x_k(t) + F'_k z'_k$, $i, j, k = 1, \ldots, n$ respectively.

Rearrangement of the expression for $D^+\tilde{\mathcal{V}}(t)$ gives

$$D^+\tilde{\mathcal{V}}(t) = \sum_{i=1}^{n} \frac{1}{H_i + h_i\frac{d\mu_i}{d\lambda}(x'_i(t))}[L_{i+1}x_{i+1}(t) - L_i x'_i(t) -$$

$$-V'_i\mu_i(x'_i(t)) + V_{i-1}y_{i-1}(t) + F_i z_i]\,[-\sigma_i(t)(L_i +$$

$$+V'_i\frac{d\mu_i}{d\lambda}(x'_i(t)) + \tilde{\sigma}_i(t)L_i + \hat{\sigma}_i(t)V'_i\frac{d\mu_i}{d\lambda}(x'_i(t))] +$$

$$+\sum_{j=2}^{n} \frac{L_j^2}{H'_j}[x'_j(t) - x_j(t)](\sigma_j(t) - \tilde{\sigma}_j(t)) -$$

$$-\tilde{\sigma}_1(t)\frac{L_1^2}{H'_1}[(x'_1(t)) - x_1(t)] +$$

$$+\sum_{k=1}^{n-1} \frac{V_k}{h'_k}[V'_k\mu_k(x'_k(t)) - V_k y_k(t) + F'_k z'_k](\sigma_k(t) - \hat{\sigma}_k(t) -$$

$$-\hat{\sigma}_n(t)\frac{V_n}{h'_n}[V'_n\mu_n(x'_n(t)) - V_n y_n(t) + F'_n z'_n].$$

But we have, for instance

$$[L_{i+1}x_{i+1}(t) - L_i x'_i(t) - V'_i\mu_i(x'_i(t)) + V_{i-1}y_{i-1}(t) + F_i z_i]\cdot$$

$$\cdot\left[-\sigma_i(t)\left(L_i + V_i'\frac{d\mu_i}{d\lambda}(x_i'(t))\right) + \tilde{\sigma}_i(t)L_i + \hat{\sigma}_i(t)V_i'\frac{d\mu_i}{d\lambda}(x_i'(t))\right] =$$

$$= \left[L_i + V_i'\frac{d\mu_i}{d\lambda}(x_i'(t))\right] |L_{i+1}x_{i+1}(t) - L_i x_i'(t) - V_i'\mu_i(x_i'(t)) +$$

$$+V_{i-1}y_{i-1}(t) + F_i z_i| + \left[\tilde{\sigma}_i(t)L_i + \hat{\sigma}_i(t)V_i'\frac{d\mu_i}{d\lambda}(x_i'(t))\right]\cdot$$

$$\cdot[L_{i+1}x_{i+1}(t) - L_i x_i'(t) - V_i'\mu_i(x_i'(t)) + V_{i-1}y_{i-1}(t) + F_i z_i] \le 0.$$

Therefore

$$D^+\tilde{\mathcal{V}}(t) \le -\frac{L_1^2}{H_1'}|(x_1'(t)) - x_1(t)| -$$

$$-\frac{V_n}{h_n'}|V_n'\mu_n(x_n'(t)) - V_n y_n(t) + F_n' z_n'| \le 0$$

and $\mathcal{V}(X(t))$ is not increasing, which results in stability of the equilibrium point.

Consider now some solution of the system which, by virtue of the fact that it starts inside the invariant set $[0,1]^{3n}$, is bounded. Consequently, its ω–limit set is not empty. On the other hand, $\tilde{\mathcal{V}}(t) = \mathcal{V}(X(t))$ is monotonically decreasing and bounded from below, hence the existence of $\lim_{t\to\infty}\tilde{\mathcal{V}}(t) = \lim_{t\to\infty}\mathcal{V}(X(t)) = \mathcal{V}_0$ follows. Consider now some point $\bar{X}$ in the ω–limit set of the considered solution; there exists $(t_k)_k$, $t_k \to \infty$ such that $X(t_k) \to \bar{X}$, $\mathcal{V}(X(t_k)) \to \mathcal{V}(\bar{X})$, the last limit follows from the continuity of $\mathcal{V}$. From the uniqueness of the limit it follows that $\mathcal{V}(\bar{X}) = \mathcal{V}_0$, hence $\mathcal{V}$ is constant on the ω–limit set; consider now the trajectory $X(t,\bar{X})$ starting in the ω–limit set; from the invariance it follows that this trajectory consists of ω–limit points only, hence $D^+\tilde{\mathcal{V}}(t) = 0$ along such trajectories. Therefore, the ω–limit set belongs to the largest invariant set contained in the set where $D^+\tilde{\mathcal{V}}(t) = 0$. (Note that, in fact, we have here the Barbašin – Krasovskii – La Salle theorem for the case of a non–smooth Liapunov function). This last set consists only of the unique equilibrium point. Indeed, on this set we have

$$x_1' - x_1 = 0$$

hence from (4.3.12b) it follows $x_1 \equiv$ const hence $x_1' \equiv$ const what gives

$$L_2 x_2 - L_1 x_1' - V_1' \mu_1(x_1') + F_1 z_1 = 0$$

From here, $x_2 \equiv$ const. On the other hand, from the equation for y_1, we have that if $x_1' \equiv$ const, then on any invariant set $y_1 \equiv$ const. Because $x_2 \equiv$ const, $x_2' \equiv x_2 \equiv$ const, hence $y_2 \equiv$ const and $x_3 \equiv$ const (as above). Proceeding by induction, we find that the only invariant set with the required properties is the unique steady state. In this way asymptotic stability of this steady state follows. This asymptotic stability is global in the sense that all solutions with physical significance i.e. belonging to the set $[0, 1]^{3n}$ tend asymptotically to the steady state.

It remains to be proven that this stability is even exponential. With respect to this we will take the same approach as in the case of steam–turbine stabilization (see Chapter 3). We prove first that this property holds in a neighbourhood of the steady–state. Indeed, the linearized system around the steady state reads:

$$H_j \frac{d\xi_j}{dt} = \frac{1}{1 + \frac{h_j}{H_j}\frac{d\mu_j}{d\lambda}(x_j^*)} \Bigg[L_{j+1}\xi_{j+1} + V_{j-1}\eta_{j-1} -$$

$$- \left(L_j + V_j' \frac{d\mu_j}{d\lambda}(x_j^*) \right) \xi_j' \Bigg] \qquad (4.3.28)$$

$$H_j' \frac{d\xi_j}{dt} = L_j(\xi_j' - \xi_j)$$

$$h_j' \frac{d\eta_j}{dt} = V_j' \frac{d\mu_j}{d\lambda}(x_j^*)\xi_j' - V_j\eta_j$$

and it is clear that this system has the same structure as (4.3.12b).

Consequently, with this system, we can associate a Liapunov function with the same structure as in (4.3.26a), namely

$$\mathcal{V}_l(\xi',\xi,\eta) = \sum_{i=1}^{n} \left|L_{i+1}\xi_{i+1} - \left(L_i + V_i'\frac{d\mu_i}{d\lambda}(x_i^*)\right)\xi_i' + V_{i-1}\eta_{i-1}\right| +$$

$$+\sum_{j=1}^{n} L_j|\xi_j' - \xi_j| + \sum_{k=1}^{n}\left|V_k'\frac{d\mu_k}{d\lambda}(x_k^*)\xi_k' - V_k\eta_k\right|$$

This will give, as before

$$D^+\tilde{\mathcal{V}}_l(t) \le -\frac{L_1^2}{H_1'}|\xi_1'(t) - \xi_1(t)| -$$

$$-\frac{V_n}{h_n'}\left|V_n'\frac{d\mu_n}{d\lambda}(x_n^*)\xi_n'(t) - V_n\eta_n(t)\right|$$

hence the linearized system is asymptotically stable i.e. exponentially stable. From the stability by the first approximation this holds also for the nonlinear system in some neighbourhood of the steady state:

$$|X(t) - X^*| \le \beta_0 e^{-\alpha t}|X(0) - X^*|$$

for $|X - X^*| \le \delta_0$. But, as it has been pointed out in the Appendix 5 of Chapter 3, the asymptotic stability means that

$$|X(t) - X^*| \le \psi(t)\varphi(|X(0) - X^*|).$$

Denoting

$$\beta_1 = \max_{|\rho|\ge\delta_0,\, \rho\in K} \frac{\varphi(\rho)}{\rho}, \qquad \tilde{\psi}(t) = \max\{\psi(t), e^{-\alpha t}\},$$

where the compact set K is defined by

$$K = \{X : -x_i^* \le x_i' - x_i^* \le 1 - x_i^*,\ -x_j^* \le x_j - x_j^* \le 1 - x_j^*,$$

$$-y_k^* \le y_k - y_k^* \le 1 - y_k^*,\ i,j,k = 1,\dots,n\}$$

we have

$$|X(t) - X^*| \le \tilde{\beta}|X(0) - X^*|\tilde{\psi}(t),$$

where $\tilde{\beta} = \max\{\beta_0, \beta_1\}$. Applying the result of Halanay, reproduced in Appendix 5 of Chapter 3, exponentially stability follows. This ends the proof.

Appendix

Here, we present the proof of the result concerning invariant sets for systems of ordinary differential equations. This result is quite important since it is based on the existence theorem on locally closed sets which is not very customary. This result belongs to Pavel and Turinici (1978) and reads

Proposition 1 *Consider the continuous mapping* $F : [t_0, t_1) \times \prod_{i=1}^{n}[a_i, b_i] \longrightarrow R^n$. *The necessary and sufficient condition that for any* $z^0 \in \prod_{i=1}^{n}[a_i, b_i]$ *there exists* $T(z_0) > 0$ *and a solution* $z : [t_0, t_0 + T(z^0)] \longrightarrow R^n$ *of the Cauchy problem*

$$z' = F(t, z), \qquad z(t_0) = z^0 \tag{A.1}$$

such that

$$z(t) \in \prod_{i=1}^{n}[a_i, b_i], \quad t_0 \leq t \leq t_0 + T(z^0)$$

is

$$F_i(t, z_1, z_2, \ldots, z_{i-1}, a_i, z_{i+1}, \ldots, z_n) \geq 0$$

$$F_i(t, z_1, z_2, \ldots, z_{i-1}, b_i, z_{i+1}, \ldots, z_n) \leq 0, \;\; i = 1, \ldots, n \tag{A.2}$$

for $t_0 \leq t < t_1$, $a_k \leq z_k \leq b_k$, $k \neq i$.

The proof of this Proposition is based on a Lemma due to M.G. Crandall (1973) which represents a generalization of Peano's existence theorem on locally closed subsets of the Euclidean space. For the sake of completeness we will give here first the result of Crandall.

Lemma 1 *Let* M *be a locally closed subset of the Euclidean space* R^n *in the sense that for any* $x \in M$ *there is some* $r > 0$ *such that* $M_r = M \cap B(x, r)$ *is closed in* R^n *where* $B(x, r)$ *is the closed ball of radius* r *and centre* $x \in R^n$. *Let* $F : [t_0, t_1) \times M \longrightarrow R^n$ *be a continuous function.*

A necessary and sufficient condition for the existence of a local solution $u : [t_0, t_0 + T(x)] \longrightarrow R^n$ *of the Cauchy problem*

$$u' = F(t, u), \qquad u(t_0) = x \tag{A.3}$$

is

$$\lim_{h \searrow 0} \frac{1}{h} d(x + hF(t, x), M) = 0 \tag{A.4}$$

for all $t_0 \leq t < t_1$ *and* $x \in M$ *where* $d(y, M)$ *is the distance from the point* $y \in R^n$ *to the set* M.

Proof The necessity part is rather easy to prove. Let (t, x) be arbitrary in $[t_0, t_1) \times M$. There exists some $T(t, x)$ and $u : [t, t + T(t, x)] \longrightarrow M$ continuously differentiable such that

$$u'(\tau) = F(\tau, u(\tau)), \quad \tau \in [t, t + T], \qquad u(t) = x$$

Also

$$\frac{1}{h} d(x + hF(t, x), M) \leq \frac{1}{h} |x + hF(t, x) - u(t + h)| =$$

$$= \left| \frac{u(t + h) - u(t)}{h} - F(t, x) \right|$$

for all $h > 0$ small enough which implies (A.4).

Conversely, let $x \in M$, $r > 0$, $\mu \geq 1$, $T(x) > 0$ be such that $M_r = M \cap B(x, r)$ is closed, $T\mu \leq r$, $|F(t, y)| \leq \mu - 1$ for all $t \in [t_0, t_0 + T]$, $y \in M_r$.

A. Let m be an arbitrary natural number. We will construct an approximate solution of (A.3) $u^m : [t_0, t_0 + T] \longrightarrow R^n$ of polygonal type as follows. Define first:

$$u_0 = u^m(t_0) = x$$

and choose the largest number $d_0 \in \left(0, \frac{1}{m}\right]$ such that

$$d(u_0 + d_0 F(t_0, u_0), M) \leq \frac{d_0}{2m}$$

$$t_1 = t_0 + d_0 \leq t_0 + T.$$

This choice is possible due to the fact that $x \in M$ and to (A.4). The first inequality above shows that there exists $u_1 \in M$ such that

$$|u_0 + d_0 F(t_0, u_0) - u_1| < d_0/m$$

Denoting

$$\Pi_0 = \frac{1}{d_0}(u_1 - u_0 - d_0 F(t_0, u_0))$$

we have

$$u_1 = u_0 + (t_1 - t_0)[F(t_0, u_0) + \Pi_0], \quad |\Pi_0| \leq \frac{1}{m}$$

and

$$|u_1 - u_0| = |u_1 - x| \leq d_0 \left(\mu - 1 + \frac{1}{m}\right) \leq T\mu \leq r$$

hence $u_1 \in M \cap B(x, r) = M_r$.

Then we construct the linear interpolation

$$u^m(t) = u_0 + (t - t_0)[F(t_0, u_0) + \Pi_0], \quad t_0 \leq t \leq t_1$$

From now on the construction goes by induction:
Assume that $u^m(t)$ has been constructed on $[t_0, t_i]$. If $t_i < t_0 + T$, choose the largest number $d_i \in (0, 1/m]$ such that

$$d(u_i + d_i F(t_i, u_i), M) \leq \frac{d_i}{2m} \tag{A.5}$$

$$t_{i+1} = t_i + d_i \leq t_0 + T \tag{A.6}$$

This choice is possible under assumption that the construction is such that $u_i \in M$ and by employing (A.4). Moreover, it follows from (A.4) that there exists $u_{i+1} \in M$ such that

$$|u_i + d_i F(t_i, u_i) - u_{i+1}| < d_i/m. \tag{A.7}$$

Denoting

$$\Pi_i = \frac{1}{d_i}(u_{i+1} - u_i - d_i F(t_i, u_i)) \tag{A.8}$$

we have

$$u_{i+1} = u_i + (t_{i+1} - t_i)[F(t_i, u_i) + \Pi_i], \quad |\Pi_i| \leq \frac{1}{m} \tag{A.9}$$

and

$$|u_{i+1} - x| \leq |u_{i+1} - u_i| + |u_i - u_{i-1}| + \ldots + |u_1 - x| \leq$$

$$\leq (d_0 + d_1 + \ldots + d_i)\left(\mu - 1 + \frac{1}{m}\right) \leq T\mu \leq r$$

hence $u_{i+1} \in M \cap B(x, r) = M_r$.

We can construct now the linear interpolation

$$u^m(t) = u_i + (t - t_i)[F(t_i, u_i) + \Pi_i],$$

$$t_i \leq t \leq t_{i+1} \tag{A.10}$$

and the induction is completed. Nevertheless, it remains to show that there is a natural number i_m such that $t_{i_m} = t_0 + T$. Indeed, if we assume by contradiction that $t_i < t_0 + T, i = 1, 2, \ldots$, this monotonically increasing and bounded sequence is convergent and we have $\lim_{i\to\infty} t_i = t_*$. By (A.9) we have

$$|u_{i+1} - u_i| \leq \mu(t_{i+1} - t_i)$$

therefore $\lim_{i\to\infty} u_i = u_*$ exists too, and $u_* \in M_r$ due to the closedness of M_r. Let $d_* \in (0, 1/m]$ small enough such that

$$d(u_* + d_* F(t_*, u_*), M) \leq \frac{d_*}{4m}.$$

It follows that for i large enough (in fact larger than some i_0) we have

$$d(u_i + d_* F(t_i, u_i), M) \leq d(u_* + d_* F(t_*, u_*), M) + \frac{d_*}{4m} \leq \frac{d_*}{2m}.$$

But d_i is the largest number in $(0, 1/m]$ satisfying (A.5); it follows from the above inequality that $d_i \geq d_*$ for all $i \geq i_0$, a contradiction to $\lim_{i\to\infty} d_i = \lim_{i\to\infty}(t_{i+1} - t_i) = 0$.

This contradiction shows that there is i_m such that $t_i \geq t_0 + T$, $i > i_m$. We can assume $t_{i_m} = t_0 + T$. In this way we have defined the polygonal line $u^m : [t_0, t_0 + T] \longrightarrow R^n$ by (A.10), $i = 0, 1, \ldots, i_m - 1$.

B We will prove some properties of the polygonal line $u^m(t)$ defined above and show that it is indeed an approximate solution of (A.3).

It is easy to show that $u^m(t) \in B(x, r)$ for all $t \in [t_0, t_0 + T]$. Indeed, for any t we can find some $i \leq i_m - 1$ for which a formula of (A.10) type is valid, hence

$$u^m(t) - x = u_i - x + (t - t_i)[F(t_i, u_i) + \Pi_i] =$$

$$= (u_i - u_{i-1}) + (u_{i-1} - u_{i-2}) + \ldots + (u_1 - x) +$$

$$+ (t - t_i)[F(t_i, u_i) + \Pi_i] =$$

$$= \sum_{k=1}^{i} (t_k - t_{k-1})[F(t_{k-1}, u_{k-1}) + \Pi_k] +$$

$$+ (t - t_i)[F(t_i, u_i) + \Pi_i] =$$

$$= \sum_{k=1}^{i} d_{k-1}[F(t_{k-1}, u_{k-1}) + \Pi_k] + (t - t_i)[F(t_i, u_i) + \Pi_i]$$

$$|u^m(t) - x| \leq \left(\sum_{k=1}^{i+1} d_{k-1}\right)\left(\mu - 1 + \frac{1}{m}\right) \leq T\mu \leq r$$

On the other hand, $u^m(t)$ is uniformly continuous, even Lipschitzian. Indeed, let $t, s \in [t_0, t_0 + T]$ and let i, j be such that $t_i \leq t \leq t_{i+1}$ and $t_j \leq s \leq t_{j+1}$; assume, for convenience that $i \geq j$. We have

$$u^m(t) - u^m(s) = u_i + (t - t_i)[F(t_i, u_i) + \Pi_i] -$$

$$- \{u_j + (s - t_j)[F(t_j, u_j) + \Pi_j]\} =$$

$$= (t - t_i)[F(t_i, u_i) + \Pi_i] + \sum_{k=j+2}^{i} (t_k - t_{k-1}) \cdot$$

$$\cdot [F(t_{k-1}, u_{k-1}) + \Pi_{k-1}] + (t_{j+1} - s)[F(t_j, u_j) + \Pi_j];$$

$$|u^m(t) - u^m(s)| \le (t - t_i + \sum_{k=j+2}^{i} (t_k - t_{k-1}) + t_{j+1} - s) \cdot$$

$$\cdot \left(\mu - 1 + \frac{1}{m} \right) \le \mu(t - s).$$

The function $u^m(t)$ can be written in the form:

$$u^m(t) = x + \sum_{j=0}^{i-1} (t_{j+1} - t_j) F(t_j, u_j) + (t - t_i) F(t_i, u_i) +$$

$$+ \sum_{j=0}^{i-1} (t_{j+1} - t_j) \Pi_j + (t - t_i) \Pi_i, \quad t_i \le t \le t_{i+1}.$$

Define the step function a^m as follows

$$a^m(s) = t_i, \quad s \in [t_i, t_{i+1}), \qquad a^m(t_0 + T) = t_0 + T.$$

Therefore

$$u^m(a^m(s)) = u^m(t_i) = u_i$$

for all $s \in [t_i, t_{i+1})$. Let

$$g^m(t) = \sum_{j=0}^{i-1} (t_{j+1} - t_j) \Pi_j + (t - t_i) \Pi_i, \quad t_i \le t \le t_{i+1}.$$

It follows that $u^m(t)$ can be written as

$$u^m(t) = x + \int_{t_0}^{t} F(a^m(s), u^m(a^m(s))) ds + g^m(t),$$

$$t \in [t_0, t_0 + T] \tag{A.11}$$

where $|g^m(t)| \le T/m$.

Since the functions $u^m(t)$, $m = 1, 2, \ldots$, are uniformly equicontinuous and belong entirely to the closed ball $B(x, r)$, as $j \to \infty$ a subsequence $(u^{m_j}(t))_j$ is uniformly convergent on $[t_0, t_0 + T]$ to a function u.

We show that $(a^{m_j}(s), u^{m_j}(a^{m_j}(s))) \to (s, u(s))$ as $j \to \infty$, uniformly on $[t_i, t_0 + T]$.

Indeed, let s be arbitrary in $[t_0, t_0+T]$ and let i be the corresponding integer for which $s \in [t_i, t_{i+1}]$, $0 \leq i \leq i_{m_j}$. It follows that $s - t_i < t_{i+1} - t_i \leq 1/m_j$, $a^{m_j}(s) = t_i$, $u^{m_j}(a^{m_j}(s)) = u_i$.

Therefore

$$|a^{m_j}(s) - s| \leq |t_i - s| \leq \frac{1}{m_j},$$

$$|u^{m_j}(a^{m_j}(s)) - u(s)| \leq |u^{m_j}(t_i) - u^{m_j}(s)| + |u^{m_j}(s) - u(s)| \leq$$

$$\leq |u^{m_j}(s) - u(s)| + \frac{\mu}{m_j}$$

and the assertion is proved.

Letting $m_j \to \infty$ in (A.11), we obtain

$$u(t) = x + \int_{t_0}^{t} F(s, u(s))ds\,, \quad t \in [t_0, t_0 + T]$$

and $u(t)$ is indeed a solution of (A.3).

But $u^{m_j}(a^{m_j}(s)) = u^{m_j}(t_i) = u_i$, hence

$$|u^{m_j}(a^{m_j}(s)) - u(s)| = |u_i - u(s)| \leq \frac{\mu}{m_j} + |u^{m_j}(s) - u(s)|$$

where $u_i \in M$. The last inequality implies $u(s) = \lim_{j\to\infty} u_i$ (taking into account that this i depends in fact on m_j). But M is closed hence $u(s) \in M$ and the Lemma is proved.

Proof of the Proposition. We will apply the Lemma with $M = \prod_{i=1}^{n} [a_i, b_i]$. By contradiction, assume first that

$$F_i(t, z_1, z_2, \ldots, z_{i-1}, a_i, z_{i+1}, \ldots, z_n) < 0$$

and denote $f_k = F_k(t, z_1, \dots, z_{i-1}, a_i, z_{i+1}, \dots, z_n)$. We have

$$\frac{1}{h} d[(z_1, \dots, z_{i-1}, a_i, z_{i+1}, \dots, z_n) + h(f_1, \dots, f_i, \dots, f_n), M] =$$
$$= |f_i| = |F_i(t, z_1, \dots, a_i, \dots, z_n)|,$$

for $t \in [t_0, t_1)$, $a_k \le z_k \le b_k$, $k \neq i$, and $h > 0$ small enough. This contradicts (A.4). Similarly we can prove that if

$$F_i(t, z_1, \dots, b_i, \dots, z_n) > 0$$

then (A.4) is again contradicted. It follows that (A.4) implies (A.2). The fact that (A.2) implies (A.4) is obvious if one realizes that in non-trivial case, in (A.2) one has to take z on the boundary of M. This equivalence of (A.2) and (A.4) in the case when $M = \prod_1^n [a_i, b_i]$ proves the Proposition.

References

BALINT, ST., STEPAN, A. (1980) *The study of the mathematical model for distillation processes in plate rectification columns.* Seminarul itinerant de ecuaţii funcţionale, aproximare şi convexitate, Timişoara, pp.269–279.

BURDESCU, D.D. (1990) *Computer Control of Chemical Processes at Craiova Chemical Plant.* Ph.D.Thesis Craiova University (in Roumanian).

BURDESCU, D.D., RĂSVAN, VL. (1990) *Stability of processes in a plate stripping tower.* in Differential Equations and Critical Theory, Pitman Research Notes in Mathematics, No. 250, pp. 51 – 57, Longman.

CRANDALL, M.G. (1973) *A generalization of Peano's existence theorem and flow invariance.* Proc.Amer.Math.Soc. **179**, pp.399–414.

FRANK–KAMENETSKII, D.A. (1987) *Diffusion and heat conduction in chemical kinetics.* Nauka, Moscow (in Russian).

FRIEDLY, J.C. (1972) *Dynamic behavior of processes.* Prentice Hall, Inc.

GOTHARD, FR., HALANAY, A., POPOV, V.M. (1968) *Stabilization of stripping process in plate rectifying columns I. The mathematical model.* Rev. Chimie **19**, *2*,pp.108–111 (in Roumanian).

KRUŽKOV, S.N., PEREGUDOV, A.N. (1990) *Cauchy problem for quasilinear parabolic equations of chemical kinetics type. Resolubility and stabilization of solutions.* Preprint A.N.S.S.S.R., Institute of structural macrokinetics Černogolovka (in Russian).

LUYBEN, W. (1973) *Process modeling, simulation and control for chemical engineers.* Mc Graw Hill.

MURGULESCU, J.G., SEGAL, E., ONCESCU, TATIANA (1981) *Introduction to physical chemistry II.2 (Chemical kinetics and catalysis).* Editura Academiei, Bucuresti (in Roumanian).

PAVEL, N., TURINICI, M. (1978) *Positive solutions of a system of ordinary differential equations.* An.St.Univ. "A.I.Cuza" Iaşi S.I.1., 63–70.

RĂSVAN, VL. (1991) *Invariant sets, Liapunov functions and singular perturbations in plate columns modeling.* Proc.Int'l Conference on Differential Equations "Equadiff 91" Barcelona.

RĂSVAN, VL. (1991) *Unifying the models of dynamics in plate columns.* An.Univ.Craiova Ses.Electrotehnică – Automatică, **15**.

REGHIS, M., STEPAN, A. (1985) *The study of distillation processes in plate rectification columns with external reflux I: The linear system of first approximation II: The stability of the non-linear system.* Seminarul de Ecuaţii Funcţionale Univ.Timişoara Papers No. 76, 77.

STEPAN, A. (1987) *Stability and controllability of differential systems describing distillation processes in plate rectifier columns.* Ph.D.Thesis, Timişoara University (in Roumanian).

VASILIEV, V.M., VOLPERT, A.I., HUDIAEV, S.I. (1973) *On the method of quasi–stationary concentrations for chemical kinetics equations.* Ž.Vych.Mat.Mat.Fyz. **13**, *3*, 683–697 (in Russian).

ZELDOVIČ, JA.B. (1938) *Proof of uniqueness of solution of mass action law equations.* Ž.Fyz.Him. **11**, *5*; 685–687.

CHAPTER 5

Stability Problems in Non – Engineering Fields

5.1 Stability of Competitive Equilibrium in Walrasian Economic Model

The Model

The state variables in the model are the prices of commodities, $p_1, \dots, p_N$, it is natural to assume that all prices are nonnegative, $p_i \geq 0$, $i = 1, \dots, N$; in general it is not excluded that some prices are zero which means that the corresponding commodities are free. The domain $P \subset R^N$ where all prices take their values is defined by

$$P = \{p \in R^N : p_i \geq 0, \ (\forall)i, \ p_j > 0, \ j \in M\}$$

where M is a given nonempty subset of $\{1, 2, \dots, N\}$, corresponding to commodities that cannot be free.

Evolution of prices is determined by the law of offer and demand; to describe the action of this law an excess demand function $E : P \longrightarrow R^N$

is considered. For a given system of prices p, the value $E_i(p)$ represents the excess of demand with respect to offer for the commodity i. The evolution of prices is described by the system of differential equations

$$\frac{dp_i}{dt} = \lambda_i E_i(p), \quad \lambda_i > 0, \quad i = 1, \dots, N \tag{5.1.1}$$

The qualitative aspect of the law of demand and offer corresponds to the fact that $E_i(p) > 0$ implies that p_i grows and $E_i(p) < 0$ implies that p_i goes down. For this model, the main problems are the existence of a system of prices corresponding to equilibrium (that is a vector p such that $E_i(p) = 0$ for all i) as well as the stability of such equilibria (in fact asymptotic stability). To answer these problems, one has to assume some properties of the excess demand function, having natural economic interpretation.

We intend to reproduce here only one result concerning stability, in order to show how such natural assumptions may be used to prove stability. We follow the book by H.Nikaido (1968), §19.2, but we expect our exposition will be selfcontained. Since we are not interested here in economic theory but in discussing interesting examples of stability problem, the reader can refer to this book for further detais.

It is assumed that functions E_i have the following properties:

1. E_i are homogeneous of degree zero, that is $E_i(\tau p) = E_i(p)$ for all $\tau > 0$. The meaning of this assumption is that the excess of the demand with respect to the offer does not depend on the price scale.

2. $\sum_1^N p_i E_i(p) = 0$ for all $p \in P$; this is the Walras Law, which we will not comment.

3. The functions E_i are C^1 and for $i \neq j$ $\dfrac{\partial E_i}{\partial p_j}(p) \geq 0$ for all $p \in P$. This property is called "*gross substitutability*" and describes the behaviour of economic agents with respect to the global system of prices: if a price of one commodity grows, the excess demand for *the other* commodities grows too.

4. There exists a real number γ such that $E_i(p) \geq \gamma$ for all i and all $p \in P$.

5. There exists $\hat{p} \in P$ with $\hat{p}_i > 0$ for all i and $E(\hat{p}) = 0$. Conditions for existence of such equilibria are result from specific research in economic theory.

As an example, let us consider the situation when $M = \{1, \dots, N\}$ and

$$E_i(p) = \frac{1}{p_i} \sum_{j=1}^{N} a_{ij} p_j - \sum_{j=1}^{N} a_{ji}, \quad a_{ij} \geq 0;$$

properties 1) and 3) are obvious; we see next that

$$\sum_{i=1}^{N} p_i E_i(p) = \sum_{i=1}^{N} \sum_{j=1}^{N} a_{ij} p_j - \sum_{i=1}^{N} \sum_{j=1}^{N} a_{ji} p_i = 0$$

and 2) is also satisfied. Property 4) is obviously true with $\gamma = -\sum_{i,j=1}^{N} a_{ij}$. Existence of an equilibrium means that the linear system

$$\sum_{j=1}^{N} a_{ij} p_j - p_i \sum_{j=1}^{N} a_{ji} = 0$$

has a solution $\hat{p}$ with $\hat{p}_i > 0$ for all i.

Properties of the Model

Proposition 5.1 *The function* $F : P \longrightarrow R$ *defined by* $F(p) = \sum_1^N \frac{1}{2\lambda_i} p_i^2$ *is a global integral for the system (5.1.1).*

Proof We have

$$\sum_1^N \frac{\partial F}{\partial p_i}(p) \lambda_i E_i(p) = \sum_1^N \frac{2p_i}{2\lambda_i} \lambda_i E_i(p) = \sum_1^N p_i E_i(p) = 0$$

Lemma 5.1 *a) For every* $k \notin M$, *we have* $E_k(p) \leq 0$ *for all* p.

b) For every k, *define* $P(k) = \{p \in P, \ p_k = 0\}$; *if* $P(k) \neq \emptyset$, *then the function* E_k *is constant on* $P(k)$.

c) If there exists $q \in P$ *such that* $E_k(q) = 0$, *then* $E_k(p) = 0$ *for all* $p \in P(k)$.

Proof a) Let $k \notin M$ and $p \in P$ with $p_j > 0$, $p_k = 0$. Define $\tilde{p}_i(t) = p_i$ for $i \neq k$, $p_k(t) = t$; $\tilde{p}(t) \in P$ for all $t \geq 0$ since $k \notin M$ and $\tilde{p}_i(t) > 0$ for $i \in M$. We have

$$E_i(\tilde{p}(t)) - E_i(\tilde{p}(0)) = \sum_1^N \frac{\partial E_i}{\partial p_l}(\tilde{q})[\tilde{p}_l(t) - \tilde{p}_l(0)] =$$

$$= \frac{\partial E_i}{\partial p_k}(\tilde{q})t \geq 0, \quad i \neq k, \quad t \geq 0.$$

We deduce that for $i \neq k$, we have $E_i(\tilde{p}(t)) \geq E_i(\tilde{p}(0))$; we have next

$$\sum_{i \neq k} \tilde{p}_i(t) E_i(\tilde{p}(t)) \geq \sum_{i \neq k} \tilde{p}_i(0) E_i(\tilde{p}_i(0)) = \sum_{i=1}^N p_i E_i(p) = 0.$$

On the other hand, $\sum_{i \neq k} \tilde{p}_i(t) E_i(\tilde{p}(t)) = -\tilde{p}_k(t) E_k(\tilde{p}(t))$ (Walras Law), hence $\tilde{p}_k(t) E(\tilde{p}_k(t)) \leq 0$ for all $t \geq 0$. Since $\tilde{p}_k(t) = t$, we deduce that $E_k(\tilde{p}(t)) \leq 0$ for all $t > 0$; for $t \to 0_+$ we deduce $E_k(p) \leq 0$.

b) Let $q \in P(k)$, $p \in P(k)$, $q_k = 0$, $p_k = 0$. Consider sequences $q^j \in P$, $p^j \in P$ with $\lim_{j \to \infty} q^j = q$, $\lim_{j \to \infty} p^j = p$, $q_i^j > 0$, $p_i^j > 0$ for $i \neq k$, $q_k^j = p_k^j = 0$. Let

$$\alpha_k^j = \min_{i \neq k} \frac{q_i^j}{p_i^j}; \quad \alpha_k^j > 0, \quad q_i^j \geq \alpha_k^j p_i^j, \quad i \neq k, \quad q_k^j = \alpha_k^j p_k^j = 0.$$

We have further

$$E_k(q^j) - E_k(\alpha_k^j p^j) = \sum_{i=1}^N \frac{\partial E_k}{\partial p_i}(\tilde{p}^j)(q_i^j - \alpha_k^j p_i^j) =$$

$$= \sum_{i \neq k} \frac{\partial E_k}{\partial p_i}(\tilde{p}^j)(q_i^j - \alpha_k^j p_i^j) \geq 0.$$

But $E_k(\alpha_k^j p^j) = E_k(p^j)$, hence $E_k(q^j) \geq E_k(p^j)$. We can interchange p^j and q^j to deduce also that $E_k(p^j) \geq E_k(q^j)$, hence $E_k(p^j) = E_k(q^j)$. For $j \to \infty$ we obtain $E_k(p) = E_k(q)$.

c) Let $E_k(q) = 0$; if $q_k = 0$ then $q \in P(k)$ and since E_k is constant on $P(k)$, we deduce $E_k(p) = 0$ for all $p \in P(k)$. Assume now $q_k > 0$ and define $\tilde{q}_i(t) = q_i$ for $i \neq k$, $\tilde{q}_k(t) = t$; we have obviously $\tilde{q}(q_k) = q$. For $0 \leq t \leq q_k$, we have $\tilde{q}_k(t) \leq q_k$, $\tilde{q}_i(t) = q_i$, $i \neq k$,

$$E_i(q) - E_i(\tilde{q}(t)) = \sum_{l=1}^{N} \frac{\partial E_i}{\partial p_l}(\hat{q})[q_l - \tilde{q}_l(t)] =$$

$$= \frac{\partial E_i}{\partial p_k}(\hat{q})[q_k - \tilde{q}_k(t)] \geq 0, \qquad i \neq k$$

hence $\sum_{i \neq k} q_i E_i(q) \geq \sum_{i \neq k} \tilde{q}_i(t) E_i(\tilde{q}(t))$. Since $E_k(q) = 0$, we have $\sum_{i \neq k} q_i E_i(q) = \sum_{i=1}^{N} q_i E_i(q) = 0$, hence $\sum_{i \neq k} \tilde{q}_i(t) E_i(\tilde{q}(t)) \leq 0$. By using Walras Law again, we deduce that $q_k(t) E_k(\tilde{q}(t)) \geq 0$, $t \cdot E_k(\tilde{q}(t)) \geq 0$ for $t > 0$, hence $E_k(\tilde{q}(t)) \geq 0$ for all $t \geq 0$. We deduce that $E_k(\tilde{q}(0)) \geq 0$ and $\tilde{q}_k(0) = 0$. From a) we deduce that $E_k(\tilde{q}(0)) \leq 0$ hence $E_k(\tilde{q}(0)) = 0$ and $\tilde{q}(0) \in P(k)$. From b) we deduce that $E_k(p) = 0$ for all $p \in P(k)$. The lemma is completely proved.

Remark 5.1 *From Lemma 5.1, c) it follows that if* $E_i(\tilde{p}) \leq 0$ *for all* i *and* $E_i(\tilde{p}) = 0$ *if* $\tilde{p}_i > 0$, *then* $E_j(\tilde{p}) = 0$ *for all* j. *By assumption 5), we have indeed* $E_i(\hat{p}) = 0$ *for all* i, *hence* $E_i(\tilde{p}) = 0$ *for such* i *for which* $\tilde{p} \in P(i)$.

Corollary 5.1 *If* $p_k(0) = 0$ *and for all* j $p_j'(t) = \lambda_j E_j(p(t))$ *for* $t > 0$, *then* $p_k(t) = 0$ *for all* $t > 0$.

Proof By assumption 5), there exists $\hat{p}$ with $E(\hat{p}) = 0$; from Lemma 5.1 c) it follows that $E_k(p) = 0$ for $p \in P(k)$. If we take $p_k(t) \equiv 0$, we have $p_k'(t) = 0 = \lambda_k E_k(p(t))$, where $p(t)$ is defined by equations (1), $i \neq k$, restricted on $P(k)$. We obtain in this way a solution of (1) and since E is C^1, this solution is unique which proves the Corollary.

Corollary 5.2 *If* $p_k(0) > 0$ *then* $p_k(t) > 0$ *for all* $t > 0$.

This is a consequence of Corollary 5.1 and of uniqueness of solutions.

Corollary 5.3 *The set* P *is invariant with respect to the flow associated with* E *and the existence of the global integral* F *shows that all solutions are bounded.*

Proposition 5.2 *All solutions of system (1) are defined on* $[0, \infty)$.

Lemma 5.2 *Let* $\tilde{p}$ *be an equilibrium; denote* $\psi(\tilde{p}, p) = \sum_1^N \tilde{p}_i E_i(p)$. *Denote by* $I(p)$ *the subset of* $\{1, \dots, N\}$ *for which* $p_i > 0$*; assume* $\psi(\tilde{p}, \cdot)$ *takes the minimal value for* $\check{p}$ *with* $I(\check{p}) = I(\tilde{p})$. *Denote*

$$\omega(\tilde{p}, p) = \min(p_i/\tilde{p}_i) \ \ \textit{for}\ i \in I(p); \qquad \omega(\tilde{p}, \check{p}) > 0 .$$

Let $S(\tilde{p}, p) = \{i : p_i > \omega(\tilde{p}, p)\tilde{p}_i\}$. *Then for every* $q \in P$ *such that for all* j, $\check{p}_j \geq q_j \geq \omega(\tilde{p}, \check{p})\tilde{p}_j$, *we have* $E_i(q) \leq 0$ *if* $i \notin I(\tilde{p})$, $E_i(q) = 0$ *if* $i \in I(p) \setminus S(\tilde{p}, \check{p})$.

Proof For $i \notin S(\tilde{p}, \check{p})$, we have $\check{p}_i = q_i = \omega(\tilde{p}, \check{p})\tilde{p}_i$ and since E_i are homogeneous of degree zero $0 = E_i(\tilde{p}) = E_i(\omega(\tilde{p}, \check{p})\tilde{p})$. We have further

$$E_i(q) = E_i(q) - E_i(\omega(\tilde{p}, \check{p})\tilde{p}) = \sum_{l=1}^{N} \frac{\partial E_i}{\partial p_l}(\tilde{q})[q_l - \omega(\tilde{p}, \check{p})\tilde{p}_l] =$$

$$= \sum_{l \neq i} \frac{\partial E_i}{\partial p_l}(\tilde{q})[q_l - \omega(\tilde{p}, \check{p})\tilde{p}_l] \geq 0$$

$$E_i(\check{p}) - E_i(q) = \sum_{l=1}^{N} \frac{\partial E_i}{\partial p_l}(\hat{q})(\check{p}_l - q_l) = \sum_{l \neq i} \frac{\partial E_i}{\partial p_l}(\hat{q})(\check{p}_l - q_l) \geq 0$$

hence $E_i(\check{p}) \geq E_i(q) \geq 0$ for $i \notin S(\tilde{p}, \check{p})$, $i \in I(\tilde{p})$.

If $i \notin I(\tilde{p}) \supset M$, it follows that $i \notin M$ and from Lemma 5.1 a) we have $E_i(q) \leq 0$. If $\check{p}$ is an equilibrium then $E_i(\check{p}) = 0$ hence $E_i(q) = 0$ for $i \in I(\tilde{p}) \setminus S(\tilde{p}, \check{p})$; it thus remains to see that $E_i(q) = 0$ for

$i \in I(\breve{p}) \setminus S(\tilde{p}, \breve{p})$ also in the case when $\breve{p}$ is not an equilibrium. Let $K(\breve{p}) = \{i : E_i(\breve{p}) > 0\}$; since $\breve{p}$ is not an equilibrium $K(\breve{p}) \neq \emptyset$ (if $E_i(\breve{p}) \leq 0$ for all i, then $\breve{p}_i$ must be an equilibrium!); $S(\tilde{p}, \breve{p}) \neq 0$ since $\breve{p}_i = \omega(\tilde{p}, \breve{p})\tilde{p}_i$ for all i implies $E_j(\breve{p}) = E_j(\omega(\tilde{p}, \breve{p})\tilde{p}) = E_j(\tilde{p}) = 0$ and $\breve{p}$ would be an equilibrium. Since E_j are continuous, we have $K(p) \supset K(\breve{p})$ for all p in a neighbourhood of $\breve{p}$, in particular for $p_i^j = \breve{p}_i$ if $i \neq j$, $p_j^j = \breve{p}_j + \delta_j$ with δ_j sufficiently small. Define $d_{ij} = E_i(\breve{p}) - E_i(p^j)$; we have

$$\sum_{i=1}^{N} d_{ij}\breve{p}_i = \sum_{i=1}^{N} \breve{p}_i E_i(\breve{p}) - \sum_{i=1}^{N} \breve{p}_i E_i(p^j) = -\sum_{i=1}^{N} \breve{p}_i E_i(p^j) =$$

$$= -\sum_{i \neq j} p_i^j E_i(p^j) - (p_j^j - \delta_j) E_j(p^j) = \delta_j E_j(p^j).$$

Therefore

$$\breve{p}_j d_{jj} = \delta_j E_j(p^j) - \sum_{i \neq j} d_{ij}\breve{p}_i$$

Since $\psi(\tilde{p}, \breve{p})$ is the minimal value for $\psi(\tilde{p}, \cdot)$, we have $\psi(\tilde{p}, p^j) \geq \psi(\tilde{p}, \breve{p})$; from the definition of ψ we deduce that $\sum_{i=1}^{N} \tilde{p}_i E_i(p^j) \geq \sum_{i=1}^{N} \tilde{p}_i E_i(\breve{p})$, hence $\sum_{i=1}^{N} \tilde{p}_i d_{ij} \leq 0$. It follows that $d_{jj}\tilde{p}_j \leq -\sum_{i \neq j} \tilde{p}_i d_{jj}$ and $d_{ij}\omega(\tilde{p}, \breve{p})\tilde{p}_j \leq -\sum_{i \neq j} \omega(\tilde{p}, \breve{p})\tilde{p}_i d_{ij}$. We have further

$$d_{jj}\left[\breve{p}_j - \omega(\tilde{p}, \breve{p})\tilde{p}_j\right] \geq \delta_j E_j(p^j) - \sum_{i \neq j} d_{ij}\breve{p}_i + \sum_{i \neq j} \omega(\tilde{p}, \breve{p})\tilde{p}_i d_{ij} =$$

$$= \delta_j E_j(p^j) - \sum_{i \neq j} d_{ij}\left[\breve{p}_i - \omega(\tilde{p}, \breve{p})\tilde{p}_i\right].$$

On the other hand, for $i \neq j$

$$d_{ij} = E_i(\breve{p}) - E_i(p^j) = \sum_{l=1}^{N} \frac{\partial E_i}{\partial p_l}(q^j)(\breve{p}_l - p_l^j) =$$

$$= -\frac{\partial E_i}{\partial p_j}(q^j)\delta_j \leq 0, \quad \delta_j > 0.$$

Since $\breve{p}_i - \omega(\tilde{p},\breve{p})\tilde{p}_i \geq 0$, we deduce that $d_{jj}[\breve{p}_j - \omega(\tilde{p},\breve{p})\tilde{p}_j] \geq \delta_j E_j(p^j) > 0$ if $j \in K(\breve{p})$ hence $\breve{p}_j - \omega(\tilde{p},\breve{p})\tilde{p}_j > 0$ for $j \in K(\breve{p})$ and it follows that $K(\breve{p}) \subset S(\tilde{p},\breve{p})$. For $i \notin S(\tilde{p},\breve{p})$ we have therefore $i \notin K(\breve{p})$, hence $E_i(\breve{p}) \leq 0$ and since $E_i(\breve{p}) \geq E_i(q) \geq 0$ for $i \notin S(\tilde{p},\breve{p})$, we deduce $E_i(q) = 0$ for $i \in I(\tilde{p}) \setminus S(\tilde{p},\breve{p})$.

Lemma 5.3 *If* $\tilde{p}$ *is an equilibrium then for every* $p \in P$, *we have* $\sum_1^N \tilde{p}_i E_i(p) \geq 0$.

Proof If $\tilde{p}$ is an equilibrium, then $\alpha\tilde{p}$ is an equilibrium for every $\alpha > 0$; it follows that we can assume $\sum_1^N \tilde{p}_i = 1$. Let $Q = \{p, p \in P, p_i > 0$ for all i for which $\tilde{p}_i > 0\}$. Denote as above $I(p) = \{i : p_i > 0\}$; we can write $Q = \{p, p \in P, I(p) \supset I(\tilde{p}) \supset M\}$. The set $Q \subset P$ has the same properties as P and we may consider the restriction of the flow on Q (Q is invariant as a consequence of Lemma 5.1). Define for $p \in Q$

$$F_i(p) = \begin{cases} \dfrac{\tilde{p}_i}{p_i}\displaystyle\sum_1^N p_j - 1 & i \in I(\tilde{p}) \\ -1 & i \notin I(\tilde{p}) \end{cases}$$

Choose $\epsilon > 0$ and define $G_i^\epsilon(p) = E_i(p) + \epsilon F_i(p)$ for $p \in Q$. The functions G_i^ϵ have properties 1) – 4); homogeneity is checked directly; then

$$\sum_1^N p_i F_i(p) = \sum_{i \in I(\tilde{p})} p_i F_i(p) - \sum_{i \notin I(\tilde{p})} p_i =$$

$$= \sum_{i \in I(\tilde{p})} p_i \left[\frac{\tilde{p}_i}{p_i}\sum_{j=1}^N p_j - 1\right] - \sum_{i \notin I(\tilde{p})} p_i =$$

$$= \sum_{i \in I(\tilde{p})} \tilde{p}_i \sum_{j=1}^N p_j - \sum_{i=1}^N p_i = \sum_{i=1}^N \tilde{p}_i \sum_{j=1}^N p_j - \sum_{i=1}^N p_i = 0$$

since $\sum_1^N \tilde{p}_i = 1$. We have next

$$\frac{\partial F_i}{\partial p_j}(p) = \frac{\tilde{p}_i}{p_i} > 0, \ j \neq i, \ i \in I(\tilde{p}), \ \frac{\partial F_i}{\partial p_j}(p) = 0,$$

and 3) is verified. Since $F_i(p) \geq -1$ for all p, we see that 4) holds as well. Let $\Sigma_N = \{p \in R^N : p_i \geq 0, \sum_1^N p_i = 1\}$, $\delta = \inf\{\sum_1^N \tilde{p}_i G_i^\epsilon(p), p \in \Sigma_N \cap Q\}$. We have

$$G_i^\epsilon(\tilde{p}) = E_i(\tilde{p}) + \epsilon F_i(\tilde{p}) = \epsilon F_i(\tilde{p})$$

since $\tilde{p}$ is an equilibrium. For $i \notin I(\tilde{p})$ we have then $G_i^\epsilon(\tilde{p}) = -\epsilon < 0$ and for $i \in I(\tilde{p})$ we have (see the definition of F_i) $G_i^\epsilon(\tilde{p}) = 0$; we deduce that $\delta \leq 0$. Consider a sequence $p^\nu \in \Sigma_N \cap Q$ such that $\lim_{\nu\to\infty} \sum_1^N \tilde{p}_i G_i^\epsilon(p^\nu) = \delta$; since Σ_N is compact, there exists a subsequence p^{ν_k} such that $\lim_{k\to\infty} p^{\nu_k} = \check{p} \in \Sigma_N \cap \bar{Q}$. If $\check{p} \in \partial Q$, there exists $j \in I(\tilde{p})$ such that $\check{p}_j = 0$; we have

$$\sum_{i=1}^N \tilde{p}_i G_i^\epsilon(p^{\nu_k}) \geq \gamma - \epsilon \sum_{i\neq j} \tilde{p}_i + \epsilon \tilde{p}_j F_j(p^{\nu_k}).$$

But $F_j(p^{\nu_k}) = (\tilde{p}_j / p_j^{\nu_k}) \sum_{l=1}^N p_l^{\nu_k} - 1$ and $\lim_{k\to\infty} F_j(p^{\nu_k}) = +\infty$, $\lim_{k\to\infty} \sum_{i=1}^N \tilde{p}_i G_i^\epsilon(p^{\nu_k}) = \infty$ contradicting $\delta \leq 0$. We deduce that $\check{p} \notin \partial Q$, hence $\check{p} \in \mathrm{Int}Q$, $\sum_{i=1}^N p_i G_i^\epsilon(\check{p}) = \delta$, and $\check{p}$ is a minimum of the function $\psi^\epsilon(\tilde{p}, \cdot)$ defined by $\psi^\epsilon(\tilde{p}, p) = \sum_1^N \tilde{p}_i G_i^\epsilon(p)$. We apply Lemma 5.2 and deduce that $G_i^\epsilon(\check{p}) = 0$ for $i \in I(p) \setminus S(\tilde{p}, \check{p})$. We have further $\check{p} \in \Sigma_N$, $\sum_1^N \check{p}_j = 1$, $F_i(\check{p}) = (\tilde{p}_i / \check{p}_i) - 1$ for $i \in I(\tilde{p})$; if $i \notin S(\tilde{p}, \check{p})$, then $\check{p}_i = \omega(\tilde{p}, \check{p})\tilde{p}_i$, $\tilde{p}_i / \check{p}_i = 1/\omega(\tilde{p}, \check{p})$.

From $\sum_{i\in I(\check{p})} \check{p}_i = 1$, $\sum_{i\in I(\tilde{p})} \tilde{p}_i = 1$, $I(\check{p}) \supset I(\tilde{p})$ we deduce $\sum_{i\in I(\tilde{p})} \check{p}_i \leq 1$; if for all $i \in I(\tilde{p})$ we had $\check{p}_i / \tilde{p}_i \geq 1$, then $\check{p}_i \geq \tilde{p}_i$ for all $i \in I(\tilde{p})$, $\sum_{i\in I(\tilde{p})} \check{p}_i \geq \sum_{i\in I(\tilde{p})} \tilde{p}_i = 1$, hence $\sum_{i\in I(\tilde{p})} \check{p}_i = 1$ and $\check{p}_i = \tilde{p}_i$ for all i, $S(\tilde{p}, \check{p}) = \emptyset$, $I(\tilde{p}) \cap S(\tilde{p}, \check{p}) = \emptyset$, $G_i^\epsilon(\check{p}) = 0$ for $i \in I(\tilde{p})$. Assume now that there exists $i \in I(\tilde{p})$ such that $\check{p}_i / \tilde{p}_i < 1$; then $\omega(\tilde{p}, \check{p}) < 1$, $F_i(\check{p}) > 0$ for $i \in I(\tilde{p}) \setminus S(\tilde{p}, \check{p})$. Since for at least one $j \in I(\tilde{p})$ we have $\check{p}_j = \omega(\tilde{p}, \check{p})\tilde{p}_j$, it is clear that we cannot have $I(\tilde{p}) \subset S(\tilde{p}, \check{p})$; then $I(\tilde{p}) \setminus S(\tilde{p}, \check{p}) \neq \emptyset$ and for $i \in I(\tilde{p}) \setminus S(\tilde{p}, \check{p})$

$$G_i^\epsilon(\check{p}) - G_i^\epsilon(\tilde{p}) = E_i(\check{p}) - E_i(\tilde{p}) + \epsilon F_i(\tilde{p}) = E_i(\check{p}) - E_i(\omega(\tilde{p}, \check{p})\tilde{p}) +$$

$$+ \epsilon F_i(\check{p}) = \sum_{l=1}^N \frac{\partial E_i}{\partial p_l}(\tilde{q}) [\check{p}_l - \omega(\tilde{p}, \check{p})\tilde{p}_l] + \epsilon F_i(\check{p}) =$$

$$= \sum_{l \neq i} \frac{\partial E_i}{\partial p_l}(\tilde{q}) [\check{p}_l - \omega(\tilde{p}, \check{p})\tilde{p}_l] + \epsilon F_i(\check{p}) \geq \epsilon F_i(\check{p}) > 0.$$

We deduce that $G_i^\epsilon(\check{p}) - G_i^\epsilon(\tilde{p}) > 0$ and since $G_i^\epsilon(\tilde{p}) = 0$ for $i \in I(\tilde{p})$, we have $G_i^\epsilon(\check{p}) > 0$ for $i \in I(\tilde{p}) \setminus S(\tilde{p}, \check{p})$ which is a contradiction since we have already seen that $G_i^\epsilon(\check{p}) = 0$ for $i \in I(\tilde{p}) \setminus S(\tilde{p}, \check{p})$. We deduce that $\check{p} = \tilde{p}$ and $G_i^\epsilon(\tilde{p}) = 0$ for all $i \in I(\tilde{p})$. It follows that $\delta = \sum_1^N \tilde{p}_i G_i^\epsilon(\tilde{p}) = 0$ and from the definition of δ, we deduce that for all $q \in Q \cap \Sigma_N$, $\sum_1^N \tilde{p}_i G_i^\epsilon(q) \geq 0$, that is $\sum_1^N \tilde{p}_i [E_i(q) + \epsilon F_i(q)] \geq 0$. For $\epsilon \to 0$ we get $\sum_1^N \tilde{p}_i E_i(q) \geq 0$ for all $q \in \Sigma_N \cap Q$.
Let now $p \in Q$; then $q = p / \left(\sum_1^N p_k\right) \in Q \cap \Sigma_N$, $E_i(q) = E_i(p)$, $\sum_1^N \tilde{p}_i E_i(p) = \sum_1^N \tilde{p}_i E_i(q) \geq 0$. Since every point in P is a limit of points in Q we deduce finally that $\sum_1^N \tilde{p}_i E_i(p) \geq 0$ for all $p \in P$ and Lemma 5.3 is proved.

Lemma 5.4 *For every equilibrium $\tilde{p}$ and any $p \in P$ which is not an equilibrium we have $\psi(\tilde{p}, p) = \sum_1^N \tilde{p}_i E_i(p) > 0$.*

Proof From Lemma 5.3 we already know that $\psi(\tilde{p}, p) \geq 0$. Let $\mathcal{E}$ be the set of all equilibria and denote by Δ the set of pairs $(\tilde{p}, p)$ minimizing ψ on $\mathcal{E} \times P$; from the proof of Lemma 5.3 we know that $\psi(\tilde{p}, p) \geq 0$ and $\psi(\tilde{p}, \tilde{p}) = 0$, hence $(\tilde{p}, \tilde{p}) \in \Delta$ and $\Delta \neq \emptyset$; we also see that the minimal value of ψ is zero. Lemma 5.4 will be proved if we show that Δ does not contain pairs $(\tilde{p}, p)$ for which p is not an equilibrium.

Denote by Γ the subset of Δ consisting of pairs $(\tilde{p}, p)$ for which p is *not* an equilibrium. We will prove that $\Gamma \neq \emptyset$ leads to contradiction. Denote by C the subset of Γ consisting of pairs $(\tilde{p}, p)$ for which $I(p) = I(\tilde{p})$. Let us show first that if $\Gamma \neq \emptyset$ then $C \neq \emptyset$. Indeed, let $(\tilde{p}, p) \in \Gamma$; since p is not an equilibrium then $K(p) = \{i : E_i(p) > 0\} \neq \emptyset$. Denote $I = I(\tilde{p}) \cap I(p)$; we have $I \supset M$ and from Lemma 5.1 we deduce that

$E_i(q) \le 0$ for $q \in P$, $i \notin I$, hence $I \supset K(p) \neq \emptyset$. Define now

$$\check{p}_i = \begin{cases} \tilde{p}_i\,, & i \in I \\ 0, & i \notin I \end{cases}, \qquad \underline{p}_i = \begin{cases} p_i\,, & i \in I \\ 0, & i \notin I \end{cases};$$

then $I(\check{p}) = I(\underline{p}) = I$ and we want to show that $(\check{p}, \underline{p}) \in C$. Since we have $\tilde{p}_i \ge \check{p}_i$ for all i and $\tilde{p}_i = \check{p}_i$ for $i \in I$, we deduce

$$E_i(\tilde{p}) - E_i(\check{p}) = \sum_{l=1}^{N} \frac{\partial E_i}{\partial p_l}(q)(\tilde{p}_l - \check{p}_l) =$$

$$= \sum_{l \notin I} \frac{\partial E_i}{\partial p_l}(q)(\tilde{p}_l - \check{p}_l) \ge 0\,, \quad i \in I$$

and $E_i(\check{p}) \le E_i(\tilde{p}) = 0$ for $i \in I$.
For $i \notin I$, we have $E_i(\check{p}) \le 0$ (Lemma 5.1) and we deduce $E_i(\check{p}) \le 0$ for all i; from here it is clear that $E_i(\check{p}) = 0$ for all i (Remark to Lemma 5.1) and $\check{p} \in \mathcal{E}$. From Lemma 5.1 c) we deduce that $E_i(p) = E_i(\underline{p}) = 0$ for $i \in I(\tilde{p}) \setminus I$. As above we see that for $i \in I$ we have $E_i(p) \ge E_i(\underline{p})$; by using Lemma 5.3

$$0 = \psi(\tilde{p}, p) = \sum_{1}^{N} \tilde{p}_i E_i(p) = \sum_{i \in I} \tilde{p}_i E_i(p) \ge \sum_{i \in I} \check{p}_i E_i(\underline{p}) =$$

$$= \sum_{1}^{N} \check{p}_i E_i(\underline{p}) = \psi(\check{p}, \underline{p}) \ge 0$$

hence we must have equality everywhere and $(\check{p}, \underline{p}) \in \Delta$. Since $\tilde{p}_i = \check{p}_i > 0$ for $i \in I$ and $\sum_{i \in I} \tilde{p}_i E_i(p) = \sum_{i \in I} \check{p}_i E_i(\underline{p})$, $E_i(p) \ge E_i(\underline{p})$ we deduce that $E_i(p) = E_i(\underline{p})$ for $i \in I$; it follows that $\underline{p}$ is not an equilibrium since for $i \in K(p) \subset I$ we have $E_i(\underline{p}) = E_i(p) > 0$. We deduce that $(\check{p}, \underline{p}) \in \Gamma$, $I(\check{p}) = I(\underline{p}) = I$ hence $(\check{p}, \underline{p}) \in C$. It remains to show that $C \neq \emptyset$ leads to contradiction.
Consider the function $S(\cdot, \cdot)$ defined on C; let $(\tilde{p}, p) \in C$ for which

$S(\tilde{p},p)$ has a minimal number of elements; $S(\tilde{p},p) \neq \emptyset$ since if $S(\tilde{p},p) = \emptyset$ then p is an equilibrium. We have

$$I = I(p) = I(\tilde{p}) \supset S(\tilde{p},p) \neq \emptyset$$

Let $\lambda = \min\{p_i/\tilde{p}_i\,,\ i \in S(\tilde{p},p)\}$, $\lambda \geq \omega = \omega(\tilde{p},p) > 0$. Define $\tilde{q}$ as

$$\tilde{q}_i = \begin{cases} \lambda\tilde{p}_i & \text{for } i \in S(\tilde{p},p) \\ \omega\tilde{p}_i & \text{for } i \notin S(\tilde{p},p) \end{cases}$$

Then $I(p) = I(\tilde{p}) = I(\tilde{q})$ and $p_i \geq \lambda\tilde{p}_i = \tilde{q}_i$ for $i \in S(\tilde{p},p)$, $p_i \geq \omega\tilde{p}_i = \tilde{q}_i$ for $i \in I(\tilde{p}) \setminus S(\tilde{p},p)$ and we deduce that $p_i \geq \tilde{q}_i$ for all i. We have also $\tilde{q}_i \geq \omega\tilde{p}_i$ for all i. From Lemma 5.2 we have $E_i(\tilde{q}) \leq 0$ for $i \notin S(\tilde{p},p)$; further $\frac{1}{\lambda}\tilde{q}_i = \tilde{p}_i$ for $i \in S(\tilde{p},p)$, $\frac{1}{\lambda}\tilde{q}_i = \frac{\omega}{\lambda}\tilde{p}_i \leq \tilde{p}_i$ for $i \notin S(\tilde{p},p)$. We deduce that

$$E_i(\tilde{q}) = E_i\left(\frac{1}{\lambda}\tilde{q}\right) \leq E_i(\tilde{p}) = 0$$

for $i \in S(\tilde{p},p)$ since

$$E_i(\tilde{q}) - E_i(\tilde{p}) = \sum_{j=1}^{N} \frac{\partial E_i}{\partial p_j}(\hat{q})\left(\frac{1}{\lambda}\tilde{q}_j - \tilde{p}_j\right) =$$

$$= \sum_{j \notin S(\tilde{p},p)} \frac{\partial E_i}{\partial p_j}(\hat{q})\left(\frac{1}{\lambda}\tilde{q}_j - \tilde{p}_j\right)$$

$i \in S(\tilde{p},p)$ and $j \notin S(\tilde{p},p)$ implies $j \neq i$, $\dfrac{\partial E_i}{\partial p_j}(\hat{q}) \geq 0$, $\frac{1}{\lambda}\tilde{q}_j - \tilde{p}_j \leq 0$ and $E_i(\tilde{q}) \leq E_i(\tilde{p}) = 0$. It follows that $E_i(\tilde{q}) \leq 0$ for all i and $\tilde{q}$ is an equilibrium. Using again Lemma 5.2 we have $E_i(p) = 0$ for $i \in I \setminus S(\tilde{p},p)$; we see next that

$$\psi(\tilde{q},p) = \sum_{i \in S(\tilde{p},p)} \tilde{q}_i E_i(p) = \lambda \sum_{i \in S(\tilde{p},p)} \tilde{p}_i E_i(p) = \lambda\psi(\tilde{p},p) = 0$$

and $(\tilde{q},p) \in C$; since $S(\tilde{q},p)$ is a proper subset of $S(\tilde{p},p)$, we contradict the minimality of $(\tilde{p},p)$ and Lemma 5.4 is proved.

Stability

Theorem 5.1 *All equilibria are stable.*

Proof Let $\tilde{p}$ be an equilibrium; define

$$V(p) = \sum_{1}^{N} \frac{1}{2\lambda_i}(p_i - \tilde{p}_i)^2$$

Consider an arbitrary solution $p(t)$ of the system $p_i' = \lambda_i E_i(p)$ and denote $\tilde{V}(t) = V(p(t))$. We have

$$\tilde{V}'(t) = \sum_{i=1}^{N} \frac{1}{\lambda_i}(p_i(t) - \tilde{p}_i)p_i'(t) = \sum_{1}^{N}(p_i(t) - \tilde{p}_i)E_i(p(t)) =$$

$$= -\sum_{1}^{N} \tilde{p}_i E_i(p(t))$$

From Lemma 5.3 we deduce that $\tilde{V}'(t) \leq 0$, hence $V(p(t)) \leq V(p(0))$ and

$$\frac{1}{2 \cdot \max_j \lambda_j} \sum_{1}^{N}(p_i(t) - \tilde{p}_i)^2 \leq \tilde{V}(t) \leq V(0) \leq$$

$$\leq \frac{1}{2 \cdot \min_j \lambda_j} \sum_{1}^{N}(p_i(0) - \tilde{p}_i)^2$$

$$|p(t) - \tilde{p}| \leq \sqrt{\frac{\max_j \lambda_j}{\min_j \lambda_j}}|p(0) - \tilde{p}|$$

and the equilibrium $\tilde{p}$ is stable.

The solutions are bounded and for every trajectory the ω–limit set is not empty. Let q be an ω–limit point for the solution $p(\cdot)$; then there exists a sequence $t_j \to \infty$ with $\lim_{j\to\infty} p(t_j) = q$, $\lim_{j\to\infty} \tilde{V}(t_j) = \lim_{j\to\infty} V(p(t_j)) = V(q)$. On the other hand, $\tilde{V}$ is a decreasing function, $\lim_{t\to\infty} \tilde{V}(t)$ exists and we deduce that $\lim_{t\to\infty} \tilde{V}(t) = V(q)$. It follows that V is constant along any trajectory in the ω–limit set of the solution $p(\cdot)$. Denote further by $p^q(\cdot)$ the solution with $p^q(0) = q$; then $V(p^q(t)) = V(q)$,

$$0 = \frac{d}{dt} V(p^q(t)) = \sum_1^N \frac{1}{\lambda_i} [p_i^q(t) - \tilde{p}_i] \lambda_i E_i(p^q(t)) =$$

$$= - \sum_1^N \tilde{p}_i E_i(p^q(t)),$$

and for $t = 0$ we deduce $\sum_1^N \tilde{p}_i E_i(q) = 0$. From Lemma 5.4 it follows that q must be an equilibrium. Therefore, we proved

Theorem 5.2 *Every solution tends to an equilibrium.*

Corollary 5.4 *Every equilibrium which is isolated on its level set for* F *is asymptotically stable on this level set.*

Remark 5.2 *The set of equilibria is convex.*

Proof Let indeed $\tilde{p}$, $\check{p}$ be equilibria, $q = \alpha\tilde{p} + \beta\check{p}$, $\alpha \geq 0$, $\beta \geq 0$, $\alpha + \beta = 1$. Assume q is not an equilibrium; from Lemma 5.4 we have $\sum_1^N \tilde{p}_i E_i(q) > 0$, $\sum_1^N \check{p}_i E_i(q) > 0$, hence

$$0 = \sum_1^N q_i E_i(q) = \sum_1^N \alpha\tilde{p}_i E_i(q) + \sum_1^N (1 - \alpha)\check{p}_i E_i(q)$$

If $\alpha = 0$ we have $0 = \sum_1^N \check{p}_i E_i(q) > 0$, a contradiction; if $\alpha > 0$ we have $0 \geq \alpha \sum_1^N \tilde{p}_i E_i(q) > 0$, again a contradiction.

Remark 5.3 *Assume* $\frac{\partial E_i}{\partial p_l}(p) > 0$ *for all* $l \neq i$ *and all* $p \in P$. *Then* $p \in P$ *implies* $p_l > 0$ *for all* l $(M = \{1, 2, \ldots, N\})$.

Proof Assume that there exists $q \in P$ and $k \in \{1, \ldots, N\}$ such that $q_k = 0$. Let $\alpha > 1$; we have

$$0 = E_k(\alpha q) - E_k(q) = \sum_1^N \frac{\partial E_i}{\partial p_l}(\tilde{q})(\alpha q_l - q_l) =$$

$$= \sum_{l \neq k}^{N} \frac{\partial E_i}{\partial p_l}(\tilde{q})(\alpha q_l - q_l)$$

Since for $j \in M$ we have $q_j > 0$, we deduce that $\alpha q_j > q_j$ and since $\frac{\partial E_k}{\partial p_j}(\tilde{q}) > 0$, we obtain $0 = \sum_{l \neq k} \frac{\partial E_k}{\partial p_l}(\tilde{q})(\alpha q_l - q_l) \geq \frac{\partial E_k}{\partial p_j}(\tilde{q})(\alpha - 1)q_j >$ 0, a contradiction.

Remark 5.4 *Assume again that* $\frac{\partial E_i}{\partial p_l}(p) > 0$ *for* $i \neq l$ *and all* $p \in P$. *Then every equilibrium is of the form* $\alpha\hat{p}$, $\alpha > 0$, *hence on a level set of* F *there is a unique equilibrium which is globally asymptotically stable on this level set.*

Proof Let $\tilde{p}$ be an equilibrium; let $\alpha = \min\{\tilde{p}_i/\hat{p}_i\,,\ i = 1, \ldots, N\}$. From the above remark we see that $\alpha > 0$. Define $S = \{i : \tilde{p}_i > \alpha\hat{p}_i\}$; from $i \notin S$ (such i exists from the definition of α) we have

$$0 = E_i(\tilde{p}) - E_i(\alpha\hat{p}) = \sum_{l=1}^{N} \frac{\partial E_i}{\partial p_l}(\breve{p})(\tilde{p}_l - \alpha\hat{p}_l) =$$

$$= \sum_{l \neq i} \frac{\partial E_i}{\partial p_l}(\breve{q})(p_l - \alpha\hat{p}_l);$$

if S is not empty $0 \geq \sum_{l \in S} \frac{\partial E_i}{\partial p_l}(\breve{q})(p_l - \alpha\hat{p}_l) > 0$, a contradiction. We deduce $S = \emptyset$ and $\tilde{p}_i = \alpha\hat{p}_i$ for all i.

5.2 Volterra Models of Interacting Species

In his book Volterra (1931), the author discusses a general model of the form

$$\beta_r \frac{dN_r}{dt} = \left(\epsilon_r \beta_r + \sum_{1}^{n} a_{sr} N_s\right) N_r\,, \tag{5.2.1}$$

$$a_{sr} = -a_{rs}\,, \quad s \neq r\,, \quad \beta_r > 0 \quad r = 1, \ldots, n$$

where the state variables N_r represent the number of individuals of the species belonging to the considered community.

If $a_{ss} < 0$ the biological community described by (5.2.1) will be called *dissipative.*

For a model of the form

$$\frac{dN_i}{dt} = N_i\left(\epsilon_i - \sum_1^n \gamma_{ij}N_j\right), \quad i = 1, \dots, n \tag{5.2.2}$$

the *dissipativity* is defined by existence of some $\alpha_j > 0$, $j = 1, \dots, n$ such that the quadratic form $\sum_{i=1}^n \sum_{j=1}^n \alpha_i \gamma_{ij} x_i x_j$ is *positive definite.* For (5.2.1) we have $\gamma_{ij} = \frac{a_{ij}}{\beta_i}$,

$$\sum_{i=1}^n \sum_{j=1}^n \beta_i \gamma_{ij} x_i x_j = \sum_{i=1}^n \sum_{j=1}^n a_{ij} x_i x_j =$$

$$= \sum_{i=1}^n a_{ii} x_i^2 < 0\,,$$

provided $a_{ii} < 0$, as it was stated.

Proposition 5.3 *For a dissipative community all solutions are bounded.*

Proof Note first that $N_i > 0$, $i = 1, \dots, n$ is an invariant set; in fact if $N_i(0) > 0$ and there is $\hat{t}$ such that $N_i(\hat{t}) \le 0_i$ then there is also $\tilde{t}$ such that $N_i(\tilde{t}) = 0$ and looking at the equations we deduce $N_i(t) \equiv 0$ a contradiction.

Now we compute

$$\frac{d}{dt}\sum_1^n \alpha_i N_i(t) = \sum_{i=1}^N \alpha_i N_i(t)\left[\epsilon_i - \sum_{j=1}^n \gamma_{ij} N_j(t)\right] =$$

$$= \sum_{i=1}^n \alpha_i \epsilon_i N_i(t) - \sum_{i=1}^N \sum_{j=1}^n \alpha_i \gamma_{ij} N_i(t) N_j(t)$$

From dissipativity we deduce existence of $\lambda > 0$ such that

$$\sum_{i=1}^{n}\sum_{j=1}^{n} \alpha_i \gamma_{ij} x_i x_j \geq \lambda \sum_{i=1}^{n} x_i^2$$

Let $\nu(t) = \max_i N_i(t)$, $\mu \geq \sum_{i=1}^{n} \alpha_i \epsilon_i$. Then

$$\sum_{i=1}^{n} \alpha_i \epsilon_i N_i(t) \leq \mu \nu(t),$$

$$\sum_{i=1}^{n}\sum_{j=1}^{n} \alpha_i \gamma_{ij} N_i(t) N_j(t) \geq \lambda \sum_{i=1}^{n} N_i^2(t) \geq \lambda \nu^2(t).$$

Let $N(t) = \sum_{i=1}^{n} \alpha_i N_i(t)$, $\alpha = \min_i \alpha_i$, $A = \sum_{i=1}^{n} \alpha_i$. Then

$$N(t) \geq \alpha \sum_{i=1}^{n} N_i(t) \geq \alpha \nu(t), \qquad N(t) \leq A\nu(t)$$

and

$$\dot{N}(t) \leq \mu \nu(t) - \lambda \nu^2(t) \leq \frac{\mu}{\alpha} N(t) - \frac{\lambda}{A^2} N^2(t)$$

By virtue of the comparison lemma, it follows that $N(t) \leq \tilde{N}(t)$ where $\tilde{N}$ is the solution to

$$\frac{d\tilde{N}}{dt} = \frac{\mu}{\alpha}\tilde{N} - \frac{\lambda}{A^2}\tilde{N}^2, \qquad \tilde{N}(0) = N(0)$$

Since $\tilde{N}(t)$ is positive and bounded for all $\tilde{N}(0) > 0$, the same is true for $N(t)$, hence for every $N_i(t)$.

Proposition 5.4 *If the community is dissipative it has a unique non-trivial equilibrium.*

Proof From (5.2.2) it is clear that we have to show that the system

$$\sum_{1}^{n} \gamma_{ij} N_j = \epsilon_i, \qquad i = 1, \ldots, n \tag{5.2.3}$$

has a unique solution. If this were not true there were some $\hat{N}_j$ not all equal to zero such that

$$\sum_1^n \gamma_{ij}\hat{N}_j = 0$$

hence

$$\sum_{i=1}^n \alpha_i \hat{N}_i \sum_{j=1}^n \gamma_{ij}\hat{N}_j = 0$$

for any $\alpha_i > 0$. But, in particular, by virtue of the property of dissipativity, we could take $\alpha_i > 0$. The contradiction thus obtained proves the assertion.

Note nevertheless that nothing was said about the biological significance of the equilibrium i.e. of the nontrivial solution of (5.2.3). It cannot be asserted that this solution is composed of nonnegative components.

Proposition 5.5 *Assume the community is dissipative and that the unique solution* $\hat{N}_1, \ldots, \hat{N}_n$ *of (5.2.3) satisfies* $\hat{N}_j > 0$ *for all* j *i.e. there is an equilibrium where all species coexist. Then the equilibrium corresponding to* $(\hat{N}_1, \ldots, \hat{N}_n)$ *is globally asymptotically stable. Moreover, this stability is exponential.*

Proof Let $x_i = \frac{N_i}{\hat{N}_i}$; we have

$$\dot{x}_i = \frac{N_i}{\hat{N}_i}\left(\epsilon_i - \sum_1^n \gamma_{ij}N_j\right) =$$

$$= x_i\left(\epsilon_i - \sum_1^n \gamma_{ij}\hat{N}_j x_j\right)$$

Taking into account (5.2.3), we have the following system for x_i

$$\dot{x}_i = -x_i \sum_1^n \gamma_{ij}\hat{N}_j(x_j - 1) \tag{5.2.4}$$

Consider the following function

$$V(x_1, \ldots, x_n) = \sum_1^n \alpha_i \hat{N}_i (x_i - 1 - \ln x_i) \tag{5.2.5}$$

where $\alpha_i > 0$ are taken from the definition of the dissipativity.

Taking into account the properties of the function $\varphi(x) = x - \ln x$, namely that $\varphi(x) \geq 1$ for all $x > 0$ with equality only for $x = 1$, it follows that $V(x_1, \ldots, x_n) \geq 0$ with equality only if $x_i = 1$ for all i.

Differentiating (5.2.5) along the solutions of (5.2.4) we have

$$\frac{d}{dt} \sum_1^n \alpha_i N_i (x_i - 1 - \ln x_i) = \sum_1^n \alpha_i \hat{N}_i (x_i - 1) \frac{dx_i}{dt} \frac{1}{x_i} =$$

$$= - \sum_{i=1}^n \alpha_i \hat{N}_i (x_i - 1) \sum_{j=1}^n \gamma_{ij} \hat{N}_j (x_j - 1) =$$

$$= - \sum_{i=1}^n \sum_{j=1}^n \alpha_i \gamma_{ij} \hat{N}_i (x_i - 1) \hat{N}_j (x_j - 1) \leq 0$$

with equality only if $x_i = 1$ for all i (the last inequality follows from dissipativity). Asymptotic stability follows from the standard Liapunov theorem (Theorem 2.4) while global asymptotic stability follows from the Barbašin – Krasovskii – La Salle argument (Corollary 2.2).

Exponential stability will be obtained as in previous cases (see Chapter 3 and 4) using the theorem presented in Appendix 5, Chapter 3. It is only necessary to show that the system is exponentially stable by the first approximation. Indeed, denote $\xi_i = x_i - 1$. We have

$$\dot{\xi}_i = -(1 + \xi_i) \sum_1^n \gamma_{ij} \hat{N}_j \xi_j$$

and the linear system for the first approximation reads

$$\dot{\xi}_i = - \sum_1^n \gamma_{ij} \hat{N}_j \xi_j , \qquad i = 1, \ldots, n. \tag{5.2.6}$$

Consider the function

$$V_L = \frac{1}{2}\sum_{1}^{n} \alpha_i \hat{N}_j \xi_i^2$$

where α_i are again taken from the definition of dissipativity. Differentiation along the solutions of (5.2.6) gives

$$\frac{dV_L}{dt} = -\sum_{i=1}^{n} \alpha_i \hat{N}_i \xi_i \sum_{j=1}^{n} \gamma_{ij} \hat{N}_j \xi_j =$$

$$= -\sum_{i=1}^{n}\sum_{j=1}^{n} \alpha_i \gamma_{ij} \hat{N}_i \xi_i \hat{N}_j \xi_j \leq 0$$

the last inequality being again a consequence of the dissipativity. Again a standard Liapunov argument gives the exponential stability of (5.2.6). This ends the proof.

It is interesting to consider also the case when not all $\hat{N}_i$, $i = 1, \ldots, n$ are positive. The result is the following:

Proposition 5.6 *If there exists j_0 such that $\hat{N}_{j_0} < 0$ and for all i we have $|\hat{N}_i| \neq 0$, then all solutions of* (5.2.2) *have the property*

$$\lim_{t\to\infty} \prod_{j,\hat{N}_j<0} \left(\frac{N_j(t)}{\hat{N}_j}\right)^{\alpha_j|\hat{N}_j|} = 0$$

(in fact exponentially), then there is at least one species for which $\hat{N}_j < 0$ tends to extinction provided the community is dissipative.

Proof Define $x_j = \frac{N_j}{|\hat{N}_j|}$, $j = 1, \ldots, n$. We have

$$\dot{x}_i = x_i \sum_{1}^{n} \gamma_{ij} |\hat{N}_j| \left(\frac{\hat{N}_j}{|\hat{N}_j|} - x_j\right). \tag{5.2.7}$$

Now we can compute

$$\sum_{1}^{n} \alpha_i |\hat{N}_i| \left(x_i - \frac{\hat{N}_i}{|\hat{N}_i|}\right) \frac{1}{x_i}\frac{dx_i}{dt} =$$

$$= -\sum_{i=1}^{n} \alpha_i |\hat{N}_i| \left(x_i - \frac{\hat{N}_i}{|\hat{N}_i|} \right) \sum_{j=1}^{n} \gamma_{ij} |\hat{N}_j| \left(x_j - \frac{\hat{N}_j}{|\hat{N}_j|} \right) =$$

$$= -\sum_{i=1}^{n} \sum_{j=1}^{n} \alpha_i \gamma_{ij} |\hat{N}_i| \left(x_i - \frac{\hat{N}_i}{|\hat{N}_i|} \right) |\hat{N}_j| \left(x_j - \frac{\hat{N}_j}{|\hat{N}_j|} \right) \leq$$

$$\leq -\lambda \sum_{i=1}^{n} |\hat{N}_i|^2 \left(x_i - \frac{\hat{N}_i}{|\hat{N}_i|} \right)^2 = -\lambda \sum_{i=1}^{n} \hat{N}_i^2 \left(1 - \frac{|\hat{N}_i|}{\hat{N}_i} x_i \right)^2 ,$$

where $\lambda > 0$ is taken from the property of dissipativity (see above). On the other hand

$$\sum_{i=1}^{n} \alpha_i |\hat{N}_i| \left(x_i - \frac{\hat{N}_i}{|\hat{N}_i|} \right) \frac{1}{x_i} \frac{dx_i}{dt} = \frac{d}{dt} \sum_{1}^{n} \alpha_i |\hat{N}_i| \left(x_i - \frac{\hat{N}_i}{|\hat{N}_i|} \ln x_i \right) =$$

$$= \frac{d}{dt} \sum_{1}^{n} \alpha_i \hat{N}_i \left(x_i \frac{|\hat{N}_i|}{\hat{N}_i} x_i - \ln x_i \right) = \frac{d}{dt} \sum_{1}^{n} \alpha_i \hat{N}_i \ln \frac{e^{\frac{|\hat{N}_i|}{\hat{N}_i} x_i}}{x_i} =$$

$$= \frac{d}{dt} \sum_{1}^{n} \ln \left(\frac{e^{\frac{|\hat{N}_i|}{\hat{N}_i} x_i}}{x_i} \right)^{\alpha_i \hat{N}_i} = \frac{d}{dt} \ln \prod_{1}^{n} \left(\frac{e^{\frac{|\hat{N}_i|}{\hat{N}_i} x_i}}{x_i} \right)^{\alpha_i \hat{N}_i} .$$

We deduce that

$$\ln \prod_{1}^{n} \left(\frac{e^{\frac{|\hat{N}_i|}{\hat{N}_i} x_i(t)}}{x_i(t)} \right)^{\alpha_i \hat{N}_i} - \ln \prod_{1}^{n} \left(\frac{e^{\frac{|\hat{N}_i|}{\hat{N}_i} x_i(0)}}{x_i(0)} \right)^{\alpha_i \hat{N}_i} \leq$$

$$\leq -\lambda \int_0^t \sum_{1}^{n} \hat{N}_i^2 \left(1 - \frac{|\hat{N}_i|}{\hat{N}_i} x_i(\tau) \right)^2 d\tau.$$

Finally we have

$$\prod_{1}^{n} \left(\frac{e^{\frac{|\hat{N}_i|}{\hat{N}_i} x_i(t)}}{x_i(t)} \right)^{\alpha_i \hat{N}_i} \leq \prod_{1}^{n} \left(\frac{e^{\frac{|\hat{N}_i|}{\hat{N}_i} x_i(0)}}{x_i(0)} \right)^{\alpha_i \hat{N}_i} .$$

$$\cdot \prod_{1}^{n} \exp\left(-\lambda \int_0^t \hat{N}_i^2 \left(1 - \frac{|\hat{N}_i|}{\hat{N}_i} x_i(\tau)\right)^2 d\tau\right). \qquad (5.2.8)$$

For each i for which $\hat{N}_i > 0$, we have $\alpha_i \hat{N}_i > 0$ and

$$\frac{1}{x_i(t)} \exp \frac{|\hat{N}_i|}{\hat{N}_i} x_i(t) = \frac{e^{x_i(t)}}{x_i(t)} . \geq 1 \qquad (5.2.9)$$

If $\hat{N}_j < 0$, then $|\hat{N}_j|/\hat{N}_j = -1$ and, taking into account (5.2.9), the left hand side of (5.2.8) will obey the following inequality

$$\prod_{1}^{n} \left(\frac{e^{\frac{|\hat{N}_i|}{\hat{N}_i} x_i}}{x_i}\right)^{\alpha_i \hat{N}_i} \geq \prod_{i, \hat{N}_i < 0} \left(\frac{e^{-x_i}}{x_i}\right)^{-\alpha_i |\hat{N}_i|} =$$

$$= \prod_{i, \hat{N}_i < 0} e^{\alpha_i |\hat{N}_i| x_i} (x_i)^{\alpha_i |\hat{N}_i|} \geq \prod_{i, \hat{N}_i < 0} (x_i)^{\alpha_i |\hat{N}_i|}.$$

On the other hand, let $\hat{N}_{j_0} < 0$. Then $1 - \left(|\hat{N}_{j_0}|/\hat{N}_{j_0}\right) x_{j_0} = 1 + x_{j_0} \geq 1$. Taking into account all what was said above, we see that inequalities (5.2.8) becomes

$$\prod_{i, \hat{N}_i < 0} [x_i(t)]^{\alpha_i |\hat{N}_i|} \leq \prod_{1}^{n} \left(\frac{e^{\frac{|\hat{N}_i|}{\hat{N}_i} x_i(0)}}{x_i(0)}\right)^{\alpha_i \hat{N}_i} .$$

$$\cdot \prod_{i \neq j_0} \exp\left(-\lambda \int_0^t \hat{N}_i^2 \left(1 - \frac{|\hat{N}_i|}{\hat{N}_i} x_i(\tau)\right)^2 d\tau\right) \cdot \exp(-\lambda \hat{N}_{j_0}^2 t)$$

and this completes the proof.

Other Properties of Dissipative Communities

1. We will discuss first some properties of the matrix $\Gamma = (\gamma_{ij})$ corresponding to a dissipative community. In such a case, we know that there exists a diagonal matrix $\mathrm{diag}(\alpha_j)$, $\alpha_j > 0$ such that

$$\frac{1}{2} [(\mathrm{diag}(\alpha_j))\Gamma + ((\mathrm{diag}(\alpha_j))\Gamma)^*] > 0.$$

Applying linear Liapunov theory (see Chapter 3, Appendix 3) it follows that $D(-\Gamma)$ is a Hurwitz matrix for every diagonal matrix D with strictly positive diagonal entries.

Let us now introduce the natural *definitions*

a) A matrix A is called *dissipative* if there exists a diagonal matrix $D > 0$ such that $DA + A^*D > 0$; it is equivalent to the fact that the quadratic form (DAx, x) is positive definite.

b) A matrix A is called D*-stable* if DA is Hurwitz for all diagonal matrices $D > 0$.

c) A matrix A is called *completely stable* if all its principal submatrices are D–stable.

In the case of biological communities, this last property means that they remain completely stable if some interactions between species disappear.

Notice that if A is Hurwitz, then A^* and A^{-1} are also Hurwitz; If A is D–stable then for every diagonal matrix $D > 0$ the matrix DA is Hurwitz, AD is also Hurwitz since $AD = D^{-1}(DA)D$, hence DA^* is Hurwitz and A^* is D–stable together with A. Since AD is Hurwitz, it follows that $D^{-1}A^{-1}$ is Hurwitz and we see that if A is D–stable then A^{-1} is also D–stable.

If A is dissipative, then there exists $D > 0$ such that $DA+A^*D < 0$; it follows that $AD^{-1}+D^{-1}A^* = D^{-1}(DA+A^*D)D^{-1} < 0$ and A^* is also dissipative. Analogously, if A is dissipative, A^{-1} is dissipative as well. It is now clear that if A is dissipative, then any principal submatrix is also dissipative and we deduce that *a dissipative matrix is completely stable.*

For a Volterra dissipative community, the matrix $(-\Gamma)$ is dissipative, hence it is also D–stable and completely stable. In fact D–stability is responsible for the stability properties of the equilibrium with coexistence of all species.

2. There is a natural way of obtaining dissipative communities. Let the resource be characterized by a parameter x and the available quantity be $K(x)$; assume that for a given species there is a specific density $f_i(x)$ of the consumption of the resource. A general equation for the biological community will be:

$$f_i(x)\frac{dN_i}{dt} = \frac{r_i N_i}{K(x)} f_i(x) \left[K(x) - \sum_{j=1}^{n} f_j(x) N_j\right].$$

Let

$$K_i = \int_X K(x) f_i(x) dx.$$

We obtain the system

$$K_i \frac{dN_i}{dt} = r_i N_i \left(K_i - \sum_1^n \alpha_{ij} N_j\right),$$

$$\alpha_{ij} = \int_X f_i(x) f_j(x) dx, \qquad i,j = 1,\dots,n.$$

Introducing a scaling

$$\tilde{N}_i = \frac{K_i N_i}{r_i}$$

we obtain the following system

$$\frac{d\tilde{N}_i}{dt} = \tilde{N}_i \left(r_i - \sum_1^n \gamma_{ij} \tilde{N}_j\right) \tag{5.2.10}$$

which is a Volterra model with $\gamma_{ij} = \alpha_{ij}\frac{r_i r_j}{K_i K_j}$. Since $\alpha_{ij} = \alpha_{ji}$ we have also $\gamma_{ij} = \gamma_{ji}$. System (5.2.10) obtained in this way is dissipative. Indeed, we have

$$\sum_{i=1}^{n}\sum_{j=1}^{n} \gamma_{ij}\xi_i\xi_j = \sum_{i=1}^{n}\sum_{j=1}^{n} \alpha_{ij}\frac{r_i}{K_i}\xi_i\frac{r_j}{K_j}\xi_j =$$

$$= \sum_{i=1}^{n}\sum_{j=1}^{n} \int_X f_i(x)\frac{r_i}{K_i}\xi_i f_j(x)\frac{r_j}{K_j}\xi_j dx =$$

$$= \int_X \left(\sum_1^n f_i(x) \frac{r_i}{K_i} \xi_i \right)^2 dx \geq 0.$$

Equality is reached only if $\sum_1^n f_i(x) \frac{r_i}{K_i} \xi_i = 0$ for all x and it is natural to assume that this is possible only if $\xi_i = 0$, $i = 1, \dots, n$.

A specific situation is the one for which

$$f_i(x) = \frac{1}{\sqrt{2\pi w_i^2}} \exp\left(-\frac{(x - x_i)^2}{2w_i^2} \right)$$

and if $w_i = w$ for all i and the distance between x_i and x_j is $|i - j|d$, then one obtains

$$\alpha_{ij} = \frac{1}{2w\sqrt{\pi}} \exp\left(-\frac{(i-j)^2 d^2}{4w^2} \right) = a^{(i-j)^2}, \quad a = \exp\left(-\frac{d^2}{4w^2} \right).$$

We consider a general situation with $\alpha_{ij} = \alpha(|i - j|)$ where $\alpha(\cdot)$ is decreasing and

$$\alpha(m) < \frac{1}{2}[\alpha(m-1) + \alpha(m+1)], \qquad m = 1, \dots, n-1.$$

Such case generates a specific dissipative community as it is clear from the fact that in this case the matrix (α_{ij}) is positive definite. The proof is given in Appendix 2.

3. Let us make some further remarks concerning the Volterra model with positive definite matrix Γ *under the assumption that there exists an equilibrium with* $\hat{N}_j > 0$ *for all* j. Let L be defined as

$$L(N_1, \dots, N_n) = \sum_1^n \left[(N - \hat{N}_i) - \hat{N}_i \ln \frac{N_i}{\hat{N}_i} \right]. \tag{5.2.11}$$

Consider the new variables.

$$\xi_i = \ln \frac{N_i}{\hat{N}_i}$$

and let

$$V(\xi_1, \dots, \xi_n) = L(\hat{N}_1 e^{\xi_1}, \dots, \hat{N}_n e^{\xi_n}) =$$

$$\sum_1^n \hat{N}_i \left[e^{\xi_i} - 1 - \xi_i\right]. \tag{5.2.12}$$

Note that $V(\xi_1, \ldots, \xi_n) \geq 0$ with equality only if $\xi_1 = \ldots = \xi_n = 0$. We have also

$$\frac{\partial V}{\partial \xi_i}(\xi_1, \ldots, \xi_n) = \hat{N}_i(e^{\xi_i} - 1), \quad i = 1, \ldots, n.$$

On the other hand, taking into account the fact that $\hat{N}_j$ are solution of (5.2.3) we can rewrite (5.2.2) as follows

$$\frac{dN_i}{dt} = N_i \sum_1^n \gamma_{ij}(\hat{N}_j - N_j), \quad i = 1, \ldots, n$$

hence

$$\frac{d\xi_i}{dt} = \frac{1}{N_i}\frac{dN_i}{dt} = \sum_1^n \gamma_{ij}\left(\hat{N}_j - \hat{N}_j e^{\xi_j}\right) =$$

$$= -\sum_{j=1}^n \gamma_{ij} \frac{\partial V}{\partial \xi_j}(\xi_1, \ldots, \xi_n).$$

It is now easy to see that

$$\frac{dV}{dt}(\xi_1, \ldots, \xi_n) = -\sum_{i=1}^n \sum_{j=1}^n \gamma_{ij} \frac{\partial V}{\partial \xi_i}(\xi_1, \ldots, \xi_n) \frac{\partial V}{\partial \xi_j}(\xi_1, \ldots, \xi_n)$$

and if $\Gamma > 0$, we have asymptotic stability of the equilibrium $\xi_1 = \xi_2 = \ldots = \xi_n = 0$, corresponding to the equilibrium $(\hat{N}_1, \ldots, \hat{N}_n)$.

It is worth to mentioning that such a property also has been pointed out in chemical kinetics for system (4.2.2) – see Chapter 4 – whose structure is very much like (5.2.2), under the same assumption of the existence of an equilibrium point with strictly positive components.

A Simple Model with Harvesting

Following Brauer and Sánchez (1975), let us now discuss a model of two species, one of which is harvested at a constant rate of E members per unit time while the second is undisturbed. The equations are

$$\frac{dN_1}{dt} = N_1(\epsilon_1 - \gamma_{11}N_1 - \gamma_{12}N_2) - E$$

$$\frac{dN_2}{dt} = N_2(\epsilon_2 - \gamma_{21}N_1 - \gamma_{22}N_2) \tag{5.2.13}$$

Assume first that without harvesting the second population goes to extinction; this is the case if $\det\Gamma = \gamma_{11}\gamma_{22} - \gamma_{12}\gamma_{21} > 0$, $\epsilon_1\gamma_{22} - \epsilon_2\gamma_{12} > 0$, $\epsilon_2\gamma_{11} - \epsilon_1\gamma_{21} < 0$ since under the above assumptions the system (5.2.3) written for the case $n = 2$ has a solution $(\hat{N}_1, \hat{N}_2)$ with $\hat{N}_1 > 0$, $\hat{N}_2 < 0$.

Assume now that without harvesting the community is dissipative; for this it suffices to assume besides $\det\Gamma > 0$ that $\gamma_{11} > 0$, $\gamma_{22} > 0$. Indeed the quadratic form

$$\alpha_1\gamma_{11}N_1^2 + (\alpha_1\gamma_{12} + \alpha_2\gamma_{21})N_1N_2 + \alpha_2\gamma_{22}N_2^2$$

is positive definite if and only if the following inequalities hold

$$\alpha_1\gamma_{11} > 0, \quad \alpha_2\gamma_{22} > 0, \quad (\alpha_1\gamma_{12} + \alpha_2\gamma_{21})^2 - 4\alpha_1\alpha_2\gamma_{11}\gamma_{22} > 0.$$

The last inequality can be written as

$$\alpha_1^2\gamma_{12}^2 - 2\alpha_1\alpha_2(\det\Gamma + \gamma_{11}\gamma_{22}) + \alpha_2^2\gamma_{21}^2 > 0.$$

It is now easy to see that the equation

$$\gamma_{12}^2\xi^2 - 2(\det\Gamma + \gamma_{11}\gamma_{22})\xi + \gamma_{21}^2 = 0$$

has two positive roots and we may choose $\xi > 0$ larger than the largest positive root; for such ξ and for $\alpha_1/\alpha_2 = \xi$ we obtain the required condition for dissipativity.

Under the dissipativity assumption we can apply Proposition 5.6 to see that $N_2(t) \to 0$ exponentially when $t \to \infty$ (in the absence of harvesting).

Let us now look for the equilibria in the presence of harvesting; we have

$$\tilde{N}_2 = \frac{1}{\gamma_{22}}(\epsilon_2 - \gamma_{21}\tilde{N}_1),$$

$$\tilde{N}_1\left[\epsilon_1 - \gamma_{11}\tilde{N}_1 - \frac{\gamma_{12}}{\gamma_{22}}(\epsilon_2 - \gamma_{21}\tilde{N}_1)\right] - E = 0.$$

If one takes into account the expressions of $\hat{N}_i$, $i = 1, 2$, from (5.2.3), the last equality becomes

$$\tilde{N}_1^2 - \hat{N}_1\tilde{N}_1 + \frac{\gamma_{22}}{\det\Gamma}E = 0.$$

We have equilibria with $\tilde{N}_1 > 0$ if

$$E < \frac{\hat{N}_1^2 \det\Gamma}{4\gamma_{22}}. \tag{5.2.14}$$

Assuming that $\epsilon_2 > 0$ then for $\gamma_{21} < 0$ there are no further conditions prohibiting $\tilde{N}_2 > 0$, while if $\gamma_{21} > 0$ we have additionally demand that

$$\frac{\epsilon_2}{\gamma_{21}} > \frac{\hat{N}_1 \pm \sqrt{\hat{N}_1^2 - (4\gamma_{22}E/\det\Gamma)}}{2},$$

which is a new condition for E.

Let us note nevertheless that the case $\gamma_{21} < 0$, $\epsilon_2 > 0$ with $\hat{N}_1 > 0$ i.e. $\epsilon_1\gamma_{22} - \epsilon_2\gamma_{12} > 0$ corresponds to $\hat{N}_2 > 0$ that is to the situation when without harvesting none of the two populations goes to extinction: an equilibrium with all (two) positive components exists and, as we already know, is asymptotically stable.

Under these conditions, inequality (5.2.14) shows that one has to limitate the harvesting in order to ensure existence of a new equilibrium and still having all the components positive.

It remains to show that this equilibrium for the case of harvesting is still asymptotically stable. We introduce the deviations

$$\xi_1 = N_1 - \tilde{N}_1, \qquad \xi_2 = N_2 - \tilde{N}_2$$

and write down the system in deviations

$$\frac{d\xi_1}{dt} = -(\xi_1 + \tilde{N}_1)(\gamma_{11}\xi_1 + \gamma_{12}\xi_2) + \frac{E}{\tilde{N}_1}\xi_1$$

$$\frac{d\xi_2}{dt} = -(\xi_2 + \tilde{N}_2)(\gamma_{21}\xi_1 + \gamma_{22}\xi_2)$$

The linear system of the first approximation is

$$\frac{d\xi_1}{dt} = \left(\frac{E}{\tilde{N}_1} - \gamma_{11}\tilde{N}_1\right)\xi_1 - \gamma_{12}\tilde{N}_1\xi_2$$

$$\frac{d\xi_2}{dt} = -\tilde{N}_2(\gamma_{21}\xi_1 + \gamma_{22}\xi_2)$$

and we have the following Hurwitz conditions

$$\frac{E}{\tilde{N}_1} - \gamma_{11}\tilde{N}_1 - \gamma_{22}\tilde{N}_2 < 0$$

$$\left(\frac{E}{\tilde{N}_1} - \gamma_{11}\tilde{N}_1\right)(-\gamma_{22}\tilde{N}_2) - \gamma_{12}\gamma_{21}\tilde{N}_1\tilde{N}_2 > 0.$$

The second inequality reads

$$-\frac{E}{\tilde{N}_1}\gamma_{22}\tilde{N}_2 + (\det\Gamma)\tilde{N}_1\tilde{N}_2 > 0$$

and taking into account that $\det\Gamma > 0$, $\gamma_{22} > 0$, $\tilde{N}_1, \tilde{N}_2 > 0$, we find

$$E < \frac{\tilde{N}_1^2 \det\Gamma}{\gamma_{22}}. \tag{5.2.15}$$

The first inequality leads to

$$E < \tilde{N}_1(\gamma_{11}\tilde{N}_1 + \gamma_{12}\tilde{N}_2)$$

but if we take into account the equilibrium equality

$$(\epsilon_1 - \gamma_{11}\tilde{N}_1 - \gamma_{12}\tilde{N}_2)\tilde{N}_1 = E$$

it becomes

$$E < \epsilon_1 \tilde{N}_1 - E \quad \text{i.e.} \quad E < \frac{\epsilon_1 \tilde{N}_1}{2}. \tag{5.2.16}$$

Notice now that if (5.2.14) holds, there exist two positive solutions for $\tilde{N}_1$, that is two equilibrium points with positive components which we denote by $(\tilde{N}_1', \tilde{N}_2')$ and $(\tilde{N}_1'', \tilde{N}_2'')$ assuming for convenience that $\tilde{N}_1' > \tilde{N}_1''$. It is obvious that by taking

$$E < \frac{\tilde{N}_1''}{2} \min\left\{\epsilon_1, \frac{\tilde{N}_1'' \det \Gamma}{2\gamma_{22}}\right\} \tag{5.2.17}$$

the inequalities (5.2.14), (5.2.15), (5.2.16) are fulfilled for both equilibrium points which are exponentially stable. Taking into account that $\tilde{N}_1' + \tilde{N}_1'' = \hat{N}_1$ and $\tilde{N}_1' \geq \tilde{N}_2'$, we have $\tilde{N}_1' \geq \hat{N}_1/2$. If (5.2.17) does not hold but we have still

$$E < \frac{\hat{N}_1}{2} \min\left\{\epsilon_1, \frac{\hat{N}_1 \det \Gamma}{2\gamma_{22}}\right\} \tag{5.2.17a}$$

the equilibrium point $(\tilde{N}_1', \tilde{N}_2')$ still possesses exponential stability.

In any case, the significance of the results is the same: in order to preserve both species the harvesting has to be limited to a certain amount.

Assume now that $\gamma_{21} > 0$ what means that the first species is "hunting" the second one. The fact that under this assumption, the second species could go to extinction shows that harvesting of the first species could save the second one. With respect to this, consider the condition guaranteeing $\tilde{N}_2 > 0$ which was found above, namely

$$\frac{\epsilon_2}{\gamma_{21}} > \frac{\hat{N}_1 \pm \sqrt{\hat{N}_1^2 - (4\gamma_{22}E/\det \Gamma)}}{2}.$$

Two cases can occur. Assume first that $\frac{\epsilon_2}{\gamma_{21}} - \frac{\hat{N}_1}{2} < 0$. Then all we can get from the above inequality is that

$$\frac{\epsilon_2}{\gamma_{21}} - \frac{\hat{N}_1}{2} > -\frac{1}{2}\sqrt{\hat{N}_1^2 - 4\gamma_{22}E/\det \Gamma}$$

which reduces to

$$\mathrm{E} < \frac{\det \Gamma}{\gamma_{22}} \cdot \frac{\epsilon_2}{\gamma_{21}} \left(\hat{\mathrm{N}}_1 - \frac{\epsilon_2}{\gamma_{21}} \right). \tag{5.2.18}$$

Another limitation of the harvesting is due to the fact that ϵ_2/γ_{21} i.e., the ratio reproduction/consumption is low for the second species. It is clear that only $(\tilde{\mathrm{N}}_1'', \tilde{\mathrm{N}}_2'')$ has positive components and by choosing E smaller than the smallest bound prescribed by (5.2.17) and (5.2.18), exponential stability is ensured. The biological situation can be described as follows: the reproducibility of the second species is weak, the first species is voracious and a limitated harvesting saves the second species from extinction but the community can be maintained only by limitating the harvesting.

Assume now that $\frac{\epsilon_2}{\gamma_{21}} - \frac{\hat{\mathrm{N}}_1}{2} > 0$. Then we have automatically

$$\frac{\epsilon_2}{\gamma_{21}} - \frac{\hat{\mathrm{N}}_1}{2} > -\frac{1}{2}\sqrt{\hat{\mathrm{N}}_1^2 - 4\gamma_{22}\mathrm{E}/\det \Gamma}$$

and if (5.2.17) holds then $(\tilde{\mathrm{N}}_1'', \tilde{\mathrm{N}}_2'')$ has positive components and is exponentially stable. We deduce that the limitation on harvesting is in this case relaxed but the stabilization of the community is still achieved at the lower equilibrium level for the harvested species.

Assume now that

$$\frac{\epsilon_2}{\gamma_{21}} - \frac{\hat{\mathrm{N}}_1}{2} > \frac{1}{2}\sqrt{\hat{\mathrm{N}}_1^2 - 4\gamma_{22}\mathrm{E}/\det \Gamma}$$

which gives

$$\mathrm{E} > \frac{\det \Gamma}{\gamma_{22}} \cdot \frac{\epsilon_2}{\gamma_{21}} \left(\hat{\mathrm{N}}_1 - \frac{\epsilon_2}{\gamma_{21}} \right) \tag{5.2.19}$$

This inequality ensures that both equilibrium points in the case of harvesting have positive components.

If E is chosen between the values prescribed by (5.2.19) and (5.2.17a), then the equilibrium point $(\tilde{\mathrm{N}}_1', \tilde{\mathrm{N}}_2')$ is exponentially stable. This corresponds to a higher level of population of the first species and to a guaranteed minimal level of harvesting. The choice of E from (5.2.17a) and (5.2.19) i.e., from

$$\frac{\det\Gamma}{\gamma_{22}}\cdot\frac{\epsilon_2}{\gamma_{21}}\left(\hat{N}_1-\frac{\epsilon_2}{\gamma_{21}}\right)<E<\frac{\hat{N}_1}{2}\min\left\{\epsilon_1,\frac{\hat{N}_1'\det\Gamma}{2\gamma_{22}}\right\} \tag{5.2.20}$$

is possible if the following condition hold.

$$\frac{\det\Gamma}{\gamma_{22}}\cdot\frac{\epsilon_2}{\gamma_{21}}\left(\hat{N}_1-\frac{\epsilon_2}{\gamma_{21}}\right)<\frac{\hat{N}_1^2}{4}\frac{\det\Gamma}{\gamma_{22}}$$

$$\frac{\det\Gamma}{\gamma_{22}}\cdot\frac{\epsilon_2}{\gamma_{21}}\left(\hat{N}_1-\frac{\epsilon_2}{\gamma_{21}}\right)<\frac{\epsilon_1\hat{N}_1}{2}$$

But the first inequality reduces to

$$\left(\frac{\hat{N}_1}{2}-\frac{\epsilon_2}{\gamma_{21}}\right)^2>0$$

while the second holds, for instance, if $\gamma_{12}>0$. In this case, it can be easily seen that $\epsilon_1>\frac{\hat{N}_1\det\Gamma}{2\gamma_{22}}$ and the second inequality follows from the first one. Consequently, (5.2.20) is replaced by

$$\frac{\det\Gamma}{\gamma_{22}}\frac{\epsilon_2}{\gamma_{21}}\left(\hat{N}_1-\frac{\epsilon_2}{\gamma_{21}}\right)<E<\frac{\hat{N}_1^2\det\Gamma}{4\gamma_{22}}. \tag{5.2.21}$$

Appendix 1
Existence of Equilibria in Walrasian Economic Model

In this Appendix we discuss some questions related to existence of equilibria in a Walrasian economic model in order to understand better the results concerning stability. The analysis will also show constraints on the set P which should be imposed in order to avoid trivial situations.

Proposition 1 *Let* $\Sigma_N=\left\{p : p_i\geq 0,\ \sum_{i=1}^N p_i=1\right\}$ *be a standard simplex. Let* $E:\Sigma_N\longrightarrow R^n$ *be continuous and satisfy the Walras law:* $\sum_{i=1}^N p_iE_i(p)=0$. *Then there exists* $\hat{p}$ *such that* $E_i(\hat{p})\leq 0$ *for all* i.

Proof Let $\Gamma = -E(\Sigma_N)$; Σ_N is compact, E is continuous hence Γ is compact. Consider a ball B such that $B \supset \Gamma$ and define $\theta : B \times \Sigma_N \longrightarrow \Sigma_N$ by

$$\theta_i(u, p) = \frac{p_i + \max(-u_i, 0)}{1 + \sum_{j=1}^{N} \max(-u_j, 0)};$$

it is clear that $\theta_i(u, p) \geq 0$ and $\sum_{i=1}^{N} \theta_i(u, p) = 1$, hence indeed $\theta(u, p) \in \Sigma_N$. Define $f : B \times \Sigma_N \longrightarrow B \times \Sigma_N$ as $f(u, p) = (-E(p), \theta(u, p))$; f is continuous and by Brouwer fixed point theorem, there exists a fixed point for f, $f(\hat{u}, \hat{p}) = (\hat{u}, \hat{p})$, that is $\hat{u} = -E(\hat{p})$, $\theta(\hat{u}, \hat{p}) = \hat{p}$.
We have $\hat{p}_i \sum_{i=1}^{N} \max(-\hat{u}_j, 0) = \max(-\hat{u}_i, 0)$ and by using Walras law, we deduce

$$\sum_{i=1}^{N} E_i(\hat{p}) \max(-\hat{u}_i, 0) = 0,$$

hence

$$\sum_{i=1}^{N} (-\hat{u}_i) \max(-\hat{u}_i, 0) = 0.$$

For all t, we have the equality $t \cdot \max(t, 0) = [\max(t, 0)]^2$ and from $\sum_{i=1}^{N} [\max(-\hat{u}_i, 0)]^2 = 0$ we deduce $\hat{u}_i \geq 0$ for all i, hence $E_i(\hat{p}) \leq 0$ for all i.
Let us note that $E_i(\hat{p}) \leq 0$ for all i implies $E_i(\hat{p}) = 0$ if $\hat{p}_i > 0$.

Proposition 2 (Uzawa) *Existence of a Walrasian equilibrium implies the Fixed Point Theorem of Brouwer.*

Proof Let $f : \Sigma_N \longrightarrow \Sigma_N$ be continuous, $f_i(p) \geq 0$, $\sum_{i=1}^{N} f_i(p) = 1$. Denote $\lambda(p) = \frac{(f(p), p)}{||p||^2}$; (notice that if $p \in \Sigma_N$ then $p \neq 0$ since $\sum_{i=1}^{N} p_i = 1$).

Denote further $\chi_i(p) = -\lambda(p) p_i + f_i(p)$; χ is continuous, Σ_N is compact, hence $\chi(\Sigma_N)$ is compact. We have

$$\sum_{i=1}^{N} p_i \chi_i(p) = -\lambda(p) ||p||^2 + \sum_{i=1}^{N} p_i f_i(p) = 0$$

hence χ satisfies Walras law. Existence of a Walrasian equilibrium means existence of $\hat{p}$ such that $\chi_i(\hat{p}) \leq 0$ for all i, and $\chi_i(\hat{p}) = 0$ if $\hat{p}_i > 0$, that is $\lambda(\hat{p})\hat{p}_i = f_i(\hat{p})$ if $\hat{p}_i > 0$. If $\hat{p}_i = 0$ from $\chi_i(\hat{p}) \leq 0$ it follows that $f_i(\hat{p}) \leq 0$ and since $f_i(\hat{p}) \geq 0$ we have $f_i(\hat{p}) = 0$, hence again $\lambda(\hat{p})\hat{p}_i = f_i(\hat{p})$. From $\lambda(\hat{p})\hat{p}_i = f_i(\hat{p})$ for all i, it follows $\lambda(\hat{p}) = 1$ hence $\hat{p}_i = f_i(\hat{p})$ for all i, and $\hat{p}$ is a fixed point for f.

The above results are beautiful, but they cannot be used in the case of our model since a model with all the required properties, with E defined on all of Σ_N will be trivial.

Theorem 1 *If* E_i *are* C^1, *homogeneous of degree zero, satisfy Walras law, and* $\frac{\partial E_i}{\partial p_j}(p) \geq 0$ *for* $i \neq j$, *and if* $E : \Sigma_N \longrightarrow R^n$, *then* $E(p) = 0$ *for all* $p \in \Sigma_N$.

To prove Theorem 1 we will need several preliminary results.

Lemma 1 *Let* $f(x) = \sum_{i=1}^{N}[\max(x_i, 0)]^2$. *Then* f *is convex, is* C^1, $\frac{\partial f}{\partial x_i}(x) = 2\max(x_i, 0)$ *and*

$$f(y) - f(x) \leq 2\sum_{i=1}^{N}(y_i - x_i)\max(y_i, 0).$$

Proof Let $\sigma(t) = [\max(t, 0)]^2$; a direct checking shows that σ is C^1 and $\sigma'(t) = 2\max(t, 0)$. From here we see that f is C^1 and we obtain the formula for the partial derivatives. Since σ' is increasing, we deduce that σ is convex, hence f is convex. The last property follows from

$$f(y) - f(x) = \sum_{i=1}^{N}\frac{\partial f}{\partial x_i}(\xi)(y_i - x_i) = 2\sum_{i=1}^{N}(y_i - x_i)\max(\xi_i, 0)$$

Lemma 2 *Consider the system* $\sum_{i=1}^{N} d_{ij}x_j = c_i$, $d_{ij} \leq 0$ *for* $i \neq j$. *Then the following statements are equivalent:*

(I) There exist $c_i > 0$, $i = 1, \ldots, N$ *such that the system admits a solution with* $x_j \geq 0$ *for all* j.

(II) For all $c_i \geq 0$, $i = 1, \ldots, N$ *the system admits a solution with* $x_j \geq 0$, $j = 1, \ldots, N$.

(III)
$$\begin{vmatrix} d_{11} & \ldots & d_{1k} \\ \vdots & & \vdots \\ d_{k1} & \ldots & d_{kk} \end{vmatrix} > 0 \text{ for all } k = 1, \ldots, N$$

Proof We will show that (I) $\Longrightarrow$ (III) $\Longrightarrow$ (II) $\Longrightarrow$ (I). To show that (I) $\Longrightarrow$ (III) we proceed by induction. For $N = 1$ the system reduces to $d_{11}x_1 = c_1$ and if there exists $c_1 > 0$ such that $x_1 \geq 0$ we must have $d_{11} > 0$ and (III) holds for k=1. Assume the statement is true for $n - 1$ and write the first equation as $d_{11}x_1 = c_1 - \sum_{j=2}^{n} d_{1j}x_j$; since $c_1 > 0$, $d_{1j} \leq 0$ for $j \geq 2$, $x_j \geq 0$, we deduce again $d_{11} > 0$. Using Gauss elimination, we obtain the equations $\sum_{j=2}^{n} d^*_{ij}x_j = c^*_i$, $i = 2, \ldots, n$, $d^*_{ij} = d_{ij} - \frac{d_{i1}d_{1j}}{d_{11}} \leq 0$ for $i \neq j$, $c^*_i = c_i - \frac{d_{ij}c_1}{d_{11}} > 0$. By induction, we deduce

$$\begin{vmatrix} d^*_{22} & \ldots & d^*_{2k} \\ \vdots & & \vdots \\ d^*_{k2} & \ldots & d^*_{kk} \end{vmatrix} > 0, \quad k = 2, \ldots, n$$

From

$$\begin{vmatrix} d_{11} & \ldots & d_{1k} \\ \vdots & & \vdots \\ d_{k1} & \ldots & d_{kk} \end{vmatrix} = \begin{vmatrix} d_{11} & d_{12} & \ldots & d_{1k} \\ 0 & d^*_{22} & \ldots & d^*_{2k} \\ \vdots & \vdots & & \vdots \\ 0 & d^*_{k2} & \ldots & d^*_{kk} \end{vmatrix}$$

and the conclusion follows.

The implication (III) $\Longrightarrow$ (II) is obtained again by induction using Gauss elimination, (II) $\Longrightarrow$ (I) is obvious.

Lemma 3 *Let* $K(p) = \{i : E_i(p) > 0\}$. *If* $K(p) \neq \emptyset$ *then*

$$\sum_{i,j \in K(p)} \frac{\partial E_i}{\partial p_j}(p)\xi_i \xi_j \leq 0$$

with equality only if $\xi = 0$.

Proof We have (Walras law) $\sum_{i=1}^N p_i E_i(p) = 0$; take the derivative with respect to p_j to obtain $\sum_{i=1}^N p_i \frac{\partial E_i}{\partial p_j}(p) + E_j(p) = 0$. On the other hand, the Euler theorem on homogeneous functions leads to $\sum_{i=1}^N p_i \frac{\partial E_j}{\partial p_i}(p) = 0$. We deduce that

$$\sum_{i=1}^N p_i \left[\frac{\partial E_i}{\partial p_j}(p) + \frac{\partial E_j}{\partial p_i}(p)\right] + E_j(p) = 0.$$

Denote $d_{ij}(p) = -\left[\frac{\partial E_i}{\partial p_j}(p) + \frac{\partial E_j}{\partial p_i}(p)\right] \leq 0$ for $i \neq j$. Since $\sum_{i=1}^N d_{ij}(p)p_i = E_j(p)$ we have $\sum_{i \in K(p)} d_{ij}(p)p_i \geq E_j(p) > 0$ for $j \in K(p)$. We can now apply Lemma 2 to deduce that the principal diagonal minors of the matrix $(d_{ij}(p))$ $i, j \in K(p)$ are strictly positive hence the corresponding quadratic form is positive definite. Since

$$\sum_{i,j \in K(p)} \frac{\partial E_i}{\partial p_j}(p)\xi_i \xi_j = -\frac{1}{2} \sum_{i,j \in K(p)} d_{ij}(p)\xi_i \xi_j$$

the conclusion of the lemma follows.

Proof of Theorem 1. Let $\theta_i(p) = \max(E_i(p), 0)$, $\phi(p) = \sum_{i=1}^N [\theta_i(p)]^2$. Function ϕ is continuous on Σ_N hence there exists a point $\hat{p}$ where ϕ attains the maximum. We shall show that $\phi(\hat{p}) = 0$, hence $\phi(p) \leq 0$ for all p, that is $\theta_i(p) = 0$ for all i, p, hence $E_i(p) \leq 0$ for all i and all p hence $E_i(p) = 0$ if $p_i > 0$. Since every point $p \in \Sigma_N$ with $p_i = 0$ is a limit of points p^k with $p_i^k > 0$, we deduce that $E_i(p) = 0$ for all i and all p. It remains thus to prove that $\phi(\hat{p}) = 0$. We modify the coordinate j of $\hat{p}$ by Δ_{pj} and denote by $\Delta_j p$ the vector with all coordinates equal to zero except the j–th one equal to Δ_{pj}.

Define $\tilde{p}^j$ by $\tilde{p}^j = \frac{\hat{p}+\Delta_j p}{1+\Delta_{pj}}$; we have $\sum_{i=1}^N \tilde{p}_i^j = 1$ and for Δ_{pj} conveniently chosen $\tilde{p}_i^j \geq 0$. Since $\phi(\hat{p})$ is the maximal value we have $\phi(\tilde{p}^j) - \phi(\hat{p}) \leq 0$. Lemma 1 leads to

$$0 \leq \phi(\hat{p}) - \phi(\tilde{p}^j) \leq 2\sum_{i=1}^N [E_i(\hat{p}) - E_i(\tilde{p}^j)] \max(E_i(\hat{p}), 0) =$$

$$= 2\sum_{i=1}^N \sum_{k=1}^N \frac{\partial E_i}{\partial p_k}(\tilde{q}^j)(\hat{p}_k - \tilde{p}_k^j) \max(E_i(\hat{p}), 0) =$$

$$= 2\sum_{i=1}^N \frac{\partial E_i}{\partial p_j}(\tilde{q}^j)(\hat{p}_j - \tilde{p}_j^j) \max(E_i(\hat{p}), 0) =$$

$$= -2\Delta_{pj} \sum_{i=1}^N \frac{\partial E_i}{\partial p_j}(\tilde{q}^j) \max(E_i(\hat{p}), 0);$$

for $\Delta_{pj} > 0$ we deduce that

$$\sum_{i=1}^N \frac{\partial E_i}{\partial p_j}(\tilde{q}^j) \max(E_i(\hat{p}), 0) \leq 0$$

and then for $\Delta_{pj} \to 0$ we obtain

$$\sum_{i=1}^N \frac{\partial E_i}{\partial p_j}(\hat{p}) \max(E_i(\hat{p}), 0) \leq 0 .$$

The Euler theorem gives

$$\sum_{j=1}^N \hat{p}_j \frac{\partial E_i}{\partial p_j}(\hat{p}) = 0 ,$$

hence we have

$$\sum_{i=1}^N \max(E_i(\hat{p}), 0) \sum_{j=1}^N \hat{p}_j \frac{\partial E_i}{\partial p_j}(\hat{p}) = 0 ,$$

that is

$$\sum_{j=1}^N \hat{p}_j \sum_{i=1}^N \frac{\partial E_i}{\partial p_j}(\hat{p}) \max(E_i(\hat{p}), 0) = 0 .$$

But $\hat{p}_j \geq 0$ and

$$\sum_{i=1}^{N} \frac{\partial E_i}{\partial p_j}(\hat{p}) \max(E_i(\hat{p}), 0) \leq 0;$$

we deduce $\sum_{i=1}^{N} \frac{\partial E_i}{\partial p_j}(\hat{p}) \max(E_i(\hat{p}), 0) = 0$ for all j with $\hat{p}_j > 0$.
Note that $\Delta_{pj} > 0$ implies that $\hat{p}_j \neq 1$; if $\hat{p}_j = 1$ then $\hat{p}_i = 0$ for $i \neq j$ and the Euler theorem gives $\frac{\partial E_i}{\partial p_j}(\hat{p}) = 0$ for all i and the above formula is valid.
Let now $i, j \in K(\hat{p})$, assuming that $K(\hat{p}) \neq \emptyset$. Let us show that if $j \in K(\hat{p})$ we must have $\hat{p}_j > 0$. Denote $d_{ij}(\hat{p}) = -\left(\frac{\partial E_i}{\partial p_j}(\hat{p}) + \frac{\partial E_j}{\partial p_i}(\hat{p})\right)$; we have $d_{ij}(\hat{p}) \leq 0$ for $i \neq j$, $\sum_{i=1}^{N} d_{ij}(\hat{p})\hat{p}_i = E_j(\hat{p})$, $j = 1, \ldots, N$ as in the proof of Lemma 3, hence $\sum_{i \in K(\hat{p})} d_{ij}(\hat{p})\hat{p}_i \geq E_j(\hat{p})$; we deduce that $\sum_{i \in K(\hat{p})} d_{ij}(\hat{p})\hat{p}_i > 0$ for $j \in K(\hat{p})$.
If $p_j = 0$ we still have $\sum_{i \in K(\hat{p}), i \neq j} d_{ij}(\hat{p})\hat{p}_i > 0$ and with $d_{ij}(\hat{p}) \leq 0$, $p_i \geq 0$ which is a contradiction.
Since for $j \in K(p)$, we have $\hat{p}_j > 0$; we deduce $\sum_{i=1}^{N} \frac{\partial E_i}{\partial p_j}(\hat{p})\theta_i(\hat{p}) = 0$ for all $j \in K(\hat{p})$, assuming that $K(\hat{p}) \neq \emptyset$. Since for $i \notin K(\hat{p})$ we have $\theta_i(\hat{p}) = 0$, such i do not contribute to the sum and we can write

$$\sum_{i \in K(\hat{p})} \frac{\partial E_i}{\partial p_j}(\hat{p})\theta_i(\hat{p}) = 0$$

hence

$$\sum_{i,j \in K(\hat{p})} \frac{\partial E_i}{\partial p_j}(\hat{p})\theta_i(\hat{p})\theta_j(\hat{p}) = 0.$$

Since $\theta_i(\hat{p}) > 0$ for $i \in K(\hat{p})$, we obtained a contradiction with Lemma 3, and we must have $K(\hat{p}) = \emptyset$, that is $E_i(\hat{p}) \leq 0$ for all i, $\theta_i(\hat{p}) = 0$ for all i, $\phi(\hat{p}) = 0$ and the proof ends.
Theorem 1 shows that in order to avoid the trivial situation we have, to demand that the set P does not contain the whole simplex Σ_N, hence P must have boundary points in Σ_N. We will state now the main assumptions for the boundary points of P.

A.1 There exist no boundary points with all coordinates strictly positive.

A.2 If $\tilde{p}$ is a boundary point for P and $p^\nu \in P$ are such that $\lim_{\nu\to\infty} p^\nu = \tilde{p}$, then at least for one i for which $\tilde{p}_i = 0$ we have

$$\limsup_{\nu\to\infty} E_i(p^\nu) > 0 .$$

Theorem 2 *Under assumptions* A.1 *and* A.2, *if* $E : P \longrightarrow R^N$ *satisfies the usual assumptions of the Walrasian model (*E *is* C^1, E_i *are homogeneous of degree zero,* $\sum_{i=1}^N p_i E_i(p) = 0$, $\frac{\partial E_i}{\partial p_j}(p) \geq 0$ *for* $i \neq j$*), then there exists at least one equilibrium in* P.

Proof Denote $Q = \{p : p \in P,\ p_i > 0 \text{ for all } i\}$. Choose $a_{ij} > 0$, $\sum_{i=1}^N a_{ij} = 1$, $j = 1, \ldots, N$ and define

$$F_i(p) = \frac{1}{p_i} \sum_{j=1}^N a_{ij} p_j - 1 .$$

Notice that: a) F_i are homogeneous of degree zero; b) $\sum_{i=1}^N p_i F_i(p) = 0$; c) if p is a boundary point for Q and $p_i = 0$, for every sequence $(q^\nu)_\nu$ with $q^\nu \in Q$, $\lim_{\nu\to\infty} q^\nu = p$ we have $\lim_{\nu\to\infty} F_i(q^\nu) = +\infty$; note that if p is a boundary point for Q, for at least one i it must be $p_i = 0$. Moreover $\frac{\partial F_i}{\partial p_j}(q) = \frac{1}{q_i} a_{ij} > 0$, $i \neq j$.
For $\epsilon > 0$, let us define $G_i^\epsilon(q) = E_i(q) + \epsilon F_i(q)$; G^ϵ has the same properties as E, and $\frac{\partial G_i}{\partial p_j}(q) > 0$ for $i \neq j$. For each boundary point p and for every i such that $p_i = 0$, we have $\limsup_{\nu\to\infty} G_i^\epsilon(q^\nu) = +\infty$ $(q^\nu \to p)$; if $p \in P$ we have $\lim_{\nu\to\infty} E_i(q^\nu) = E_i(p)$, $\lim_{\nu\to\infty} F_i(q^\nu) = +\infty$ and if $p \notin P$ then p is a boundary point for P and $\limsup_{\nu\to\infty} E_i(q^\nu) > 0$, $\lim_{\nu\to\infty} F_i(q^\nu) = +\infty$. Denote $\eta^\epsilon(q) = \max_i G_i^\epsilon(q)$, $q \in Q$. From Walras law, we see that we cannot have $G_i^\epsilon(q) > 0$ for all i, hence $\eta^\epsilon(q) \geq 0$. Let $\delta^\epsilon = \inf\{\eta^\epsilon(q), q \in Q \cap \Sigma_N\}$ and choose a sequence $q^{\nu,\epsilon} \in Q \cap \Sigma_N$ with $\lim_{\nu\to\infty} \eta^\epsilon(q^{\nu,\epsilon}) = \delta^\epsilon$. Since $q^{\nu,\epsilon} \in \Sigma_N$ and Σ_N is compact we

can take a subsequence converging to a point $\hat{q}^\epsilon \in \Sigma_N$; if $\hat{q}^\epsilon \in Q$ then $\eta^\epsilon(\hat{q}^\epsilon) = \lim_{j\to\infty} \eta^\epsilon(q^{\nu_j,\epsilon}) = \delta^\epsilon$ and $\hat{q}^\epsilon$ is a minimum point. If $\hat{q}^\epsilon \notin Q$ it must be a boundary point and $\limsup_{j\to\infty} \eta^\epsilon(q^{\nu_j,\epsilon}) = +\infty$ contradicting $\lim_{j\to\infty} \eta^\epsilon(q^{\nu_j,\epsilon}) = \delta^\epsilon$. We prove now that $\eta^\epsilon(\hat{q}^\epsilon) = 0$. If $\eta^\epsilon(\hat{q}^\epsilon) > 0$ there exists j such that $G_j^\epsilon(\hat{q}^\epsilon) > 0$, and by Walras law there exist i such that $\hat{q}_i^\epsilon G_i^\epsilon(\hat{q}_i^\epsilon) < 0$. Since $\hat{q}_i^\epsilon > 0$ (definition of Q) we must have $G_i^\epsilon(\hat{q}^\epsilon) < 0$ and we may find a new point $\tilde{q}^\epsilon \in Q$ with $G_i^\epsilon(\tilde{q}^\epsilon) < 0$, $\tilde{q}_i^\epsilon < \hat{q}_i^\epsilon$, $\tilde{q}_j^\epsilon = \hat{q}_j^\epsilon$, $j \neq i$. For $k \neq i$

$$G_k^\epsilon(\tilde{q}^\epsilon) - G_k^\epsilon(\hat{q}^\epsilon) = \sum_{l=1}^{N} \frac{\partial G_k^\epsilon}{\partial p_l}(q^\epsilon)(\tilde{q}_l^\epsilon - \hat{q}_l^\epsilon) =$$

$$= \frac{\partial G_k}{\partial p_i}(q^\epsilon)(\tilde{q}_i^\epsilon - \hat{q}_i^\epsilon) < 0;$$

it follows that $\eta^\epsilon(\tilde{q}^\epsilon) < \eta^\epsilon(\hat{q}^\epsilon)$ a contradiction since $\hat{q}^\epsilon$ is a minimum point.

Consider now a sequence $\epsilon_\nu \to 0$; we obtain a corresponding sequence $\hat{q}^{\epsilon_\nu}$ with $\eta^{\epsilon_\nu}(\hat{q}^{\epsilon_\nu}) = 0$. Since $\hat{q}^{\epsilon_\nu} \in Q \cap \Sigma_N$ and Σ_N is compact there exists a subsequence $\hat{q}^{\epsilon_{\nu_j}}$ such that $\lim_{j\to\infty} \hat{q}^{\epsilon_{\nu_j}} = \hat{p}$ with $\hat{p} \in \Sigma_N$. If $\hat{p} \in Q$ we have

$$0 = \eta^{\epsilon_{\nu_j}}(\hat{q}^{\epsilon_{\nu_j}}) \geq G_i^{\epsilon_{\nu_j}}(\hat{q}^{\epsilon_{\nu_j}}) = E_i(\hat{q}^{\epsilon_{\nu_j}}) + \epsilon_{\nu_j} F_i(\hat{q}^{\epsilon_{\nu_j}}) \geq$$

$$\geq E_i(\hat{q}^{\epsilon_{\nu_j}}) - \epsilon_{\nu_j},$$

that is $\epsilon_{\nu_j} \geq E_i(\hat{q}^{\epsilon_{\nu_j}})$ and for $j \to \infty$ we deduce $E_i(\hat{p}) \leq 0$ for all i and $\hat{p}$ is an equilibrium in $Q \cap \Sigma_N$ ($\hat{p}_i > 0$ for all i since $\hat{p} \in Q$). If $\hat{p}$ is not in Q it must be in P. If $\hat{p}$ were not in P, it were a boundary point for P; but $E_i(\hat{q}^{\epsilon_{\nu_j}}) \leq \epsilon_{\nu_j}$ implies $\lim_{j\to\infty} E_i(\hat{q}^{\epsilon_{\nu_j}}) \leq 0$ while if $\hat{p}$ is a boundary point for P, $\limsup_{j\to\infty} E_i(\hat{q}^{\epsilon_{\nu_j}}) > 0$. In this way existence of an equilibrium in P is proved.

Appendix 2

Theorem 1 *Consider a matrix* $A = (\alpha_{ij})$, $\alpha_{ij} = \alpha(|i-j|)$, $\alpha(m) < \frac{1}{2}[\alpha(m-1)+\alpha(m+1)]$, $m = 1,\dots,n-1$, $\alpha(\cdot)$ *decreasing. Then* $A > 0$.

Proof Denote $\alpha(m) = \alpha_m$ and assume $\alpha(0) = 1$. Associate with A the matrix $C(A)$ defined by

$$C(A) = \begin{pmatrix} 1 & \alpha_1 & \dots & \alpha_{n-1} & \alpha_n & \alpha_{n-1} & \dots & \alpha_1 \\ \alpha_1 & 1 & \dots & \alpha_{n-2} & \alpha_{n-1} & \alpha_n & \dots & \alpha_2 \\ \vdots & \vdots & & \vdots & \vdots & \vdots & & \vdots \\ \alpha_{n-1} & \alpha_{n-2} & \dots & 1 & \alpha_1 & \alpha_2 & \dots & \alpha_n \\ \alpha_n & \alpha_{n-1} & \dots & \alpha_1 & 1 & \alpha_1 & \dots & \alpha_{n-1} \\ \alpha_{n-1} & \alpha_{n-2} & \dots & \alpha_2 & \alpha_1 & 1 & \dots & \alpha_{n-2} \\ \vdots & \vdots & & \vdots & \vdots & \vdots & & \vdots \\ \alpha_1 & \alpha_2 & \dots & \alpha_n & \alpha_{n-1} & \alpha_{n-2} & \dots & 1 \end{pmatrix}$$

The eigenvalues of $C(A)$ are (R.Bellman, 1960, ch.12, §15):

$$\mu_k = 1 + 2\sum_{j=1}^{n-1} \alpha_j \cos j\frac{k\pi}{n} + (-1)^k \alpha_n\,, \quad k = 0,1,\dots,2n-1\,,$$

$$\mu_k = \mu_{2n-k}\,, \quad k = 1,\dots,n-1\,.$$

Our assumptions read

$$1 > \alpha_1 > \dots > \alpha_n > 0\,, \qquad \alpha_m < \frac{1}{2}(\alpha_{m-1} + \alpha_{m+1})\,,$$

$$m = 1,\dots,n-1$$

Let Ω be the set of vectors $(\alpha_1,\dots,\alpha_n)$ satisfying the above constraints and $\bar{\Omega}$ the closure of Ω. We will prove that

$$\min_{\bar{\Omega},k} \mu_k = 0\,, \qquad \inf_{\Omega}\left\{\min_k \mu_k\right\} = 0$$

and the infimum is not a minimum; it will follow that $C(A) > 0$.

Define

$$\alpha_n = x_n\,,\ \alpha_{n-1} = x_{n-1} + x_n\,,\ \ldots,$$

$$\alpha_m = x_m + 2x_{m+1} + \ldots + (n-m)x_{n-1} + x_n\,,\ \ldots$$

$$\ldots,\ \alpha_1 = x_1 + 2x_2 + \ldots + (n-1)x_{n-1} + x_n\,,$$

$$1 = \alpha_0 = x_0 + 2x_1 + \ldots + nx_{n-1} + x_n$$

Then $\alpha_m \leq \frac{1}{2}(\alpha_{m-1} + \alpha_{m+1})$, $m = 1, \ldots, n-1$, imply $x_{m-1} \geq 0$, $m = 1, \ldots, n-1$; from $\alpha_{n-1} \geq \alpha_n \geq 0$ we deduce $x_{n-1} \geq 0$, $x_n \geq 0$.

The eigenvalue μ_k is linear affine with respect to $\alpha_1, \ldots, \alpha_n$ hence with respect to $x_0, \ldots, x_n$. The minimum will be reached in one of the points

$$V_0 = [1, 0, \ldots, 0, 0]\,,\ V_1 = [0, \frac{1}{2}, \ldots, 0]\,, \ldots$$

$$\ldots,\ V_{n-1} = [0, 0, \ldots, \frac{1}{n}, 0]\,,\ V_n = [0, \ldots, 0, 1]$$

We have

$$\mu_k(V_m) = 1 + 2\sum_{j=1}^{m} \frac{m+1-j}{m+1} \cos j\frac{k\pi}{n} = 1 + \frac{2}{m+1} Y_m(\varphi_k)\,,$$

$$Y_m(\varphi) = m\cos\varphi + (m-1)\cos 2\varphi + \ldots + \cos m\varphi\,,\quad \varphi_k = \frac{k\pi}{n}$$

By induction, it is easy to check that

$$Y_m(\varphi) = \frac{1}{2}\left[\frac{\cos\varphi - \cos(m+1)\varphi}{1-\cos\varphi} - m\right],\quad \varphi \neq 0, \pm 2\pi, \ldots$$

We deduce that

$$\mu_k(V_m) = \frac{1}{m+1}\,\frac{1-\cos(m+1)\varphi_k}{1-\cos\varphi_k}$$

and we see that for $m = 1, \ldots, n-2$ we have $\mu_k(V_m) > 0$ and also

$$\mu_k(V_{n-1}) = \begin{cases} 0 & \text{if } k \text{ is even} \\ \dfrac{2}{n(1-\cos\varphi_k)} > 0 & \text{if } k \text{ is odd} \end{cases}$$

We have further

$$\mu_k(V_n) = 1 + 2\sum_{j=1}^{n-1} \cos j\varphi_k + (-1)^k = \frac{\sin n - \frac{1}{2}\varphi_k}{\sin\frac{\varphi_k}{2}} + (-1)^k = 0$$

for all $k = 1, \ldots, n$.

The analysis shows also that none of the points where the value $\mu_k = 0$ is reached belongs to Ω, hence in Ω we have $\mu_k > 0$ for all k.

By using the Sturm separation theorem (Bellman, 1960, ch.7 §8) we deduce for the eigenvalues of the symmetric matrix A – a principal submatrix in $C(A)$ – that $\min_k \mu_k \le \lambda_i \le \max_k \mu_k$, hence all eigenvalues λ_i are positive and $A > 0$.

Examples of the situation described in the theorem are $\alpha_{ij} = a^{|i-j|}$, $0 < a < 1$; $\alpha_{ij} = (1 + |i-j|^p)^{-1}$, $p > 0$; $\alpha_{ij} = a^{(i-j)^2}$ with α not too close to 1.

This theorem was proved by D.O.Logofet (1975).

References

BELLMAN,R. (1960) *Introduction to Matrix Analysis.* Mc Graw – Hill.

BRAUER,F., SÁNCHEZ,D.A. (1975) *Some models for population growth with harvesting.* Int'l Conference on Differential Equations edited by H.Antosiewicz, pp.53–64. Academic Press.

LOGOFET,D.O. (1975) *On stability of a class of matrices appearing in the mathematical theory of biological communities.* Dokl.A.N.S.S.S.R. **221**, *6*, 1272–1275 (in Russian).

NIKAIDO,H. (1968) *Convex structures and economic theory.* Academic Press.

VOLTERRA,V. (1931) *Leçons sur la théorie mathématique de la lutte pour la vie.* Gauthier - Villars.

Index